INTERNATIONAL ASTRONOMICAL UNION
UNION ASTRONOMIQUE INTERNATIONALE

SYMPOSIUM No. 42

HELD AT ST. ANDREWS, FIFE, SCOTLAND,
11–13 AUGUST 1970

WHITE DWARFS

EDITED BY

W. J. LUYTEN

University of Minnesota, Minneapolis, Minn., U.S.A.

D. REIDEL PUBLISHING COMPANY
DORDRECHT-HOLLAND

1971

Published on behalf of
the International Astronomical Union
by
D. Reidel Publishing Company, Dordrecht, Holland

Library of Congress Catalog Card Number 75–146966

ISBN-13:978-94-010-3083-0   e-ISBN-13:978-94-010-3081-6
DOI: 10.1007/978-94-010-3081-6

WHITE DWARFS

INTRODUCTION

When the Executive Committee of the International Astronomical Union asked me, in 1968, to organize a Symposium on White Dwarfs it became evident that members of at least four Commissions of the IAU should participate, and that the most opportune place, and time to hold such a Symposium would be somewhere in the British Isles and just preceding the Fourteenth General Assembly at Brighton in August, 1970.

After a preliminary correspondence with Dr. D. W. N. Stibbs it was decided to accept his invitation to hold the Symposium at St. Andrews University, Scotland, while the dates 11–13 August 1970 were selected. I am sure I speak for all the participants of the Symposium when I express our deep gratitude to Dr. Stibbs for the admirable way in which he planned and organized this meeting down to the smallest details – the success of the conference is due in no small measure to his continuous and painstaking efforts.

An organizing committee was formed, consisting of O. J. Eggen (Photometry), J. L. Greenstein (Spectroscopy), A. G. Masevitch (Stellar Interiors), V. Weidemann (Stellar Atmospheres), D. W. N. Stibbs (local chairman), with myself (Proper Motions and Parallaxes) as overall chairman.

Since the main purpose of such a Symposium is to have a full and extensive discussion of the subject, it was decided to limit the number of participants to 24–30, to have three morning sessions, each featuring two longer (30–40 minutes) summarizing papers, followed by discussion, and devote the afternoons to having a number of shorter, contributed papers, generally five to six, each also followed by discussion.

This was probably the first time that observational and theoretical astronomers in this field were brought together and were forced to try and understand each other's language. Whether we succeeded only time will tell.

We are greatly indebted to the International Astronomical Union for a grant which helped to defray the expenses of the Symposium.

W. J. LUYTEN

OPENING REMARKS

W. J. Luyten (Chairman of the Organizing Committee) – As I am the one who was asked by the Executive Committee of the IAU to organize this Symposium – No. 42, on White Dwarfs – it gives me great pleasure to welcome you all on this occasion.

First of all I should like – and I am sure I am speaking for all of you – to express our deep appreciation to the Court of the University of St. Andrews for their permission to hold our meetings here and especially for their generosity in making all the detailed arrangements.

And now I should like to introduce Professor Norman Gash, Vice-Principal of the University, who will formally open our proceedings.

N. Gash (Vice-Principal of the University of St. Andrews.) – Ladies and Gentlemen: My task this morning is simple, and I think I can make it brief. It is, of course, merely to welcome you to St. Andrews, to the town and to the University. It is always a pleasure for us, as I suppose for most Universities, to act as host to such a distinguished gathering of international scholars as you are. I think we should always remember that Universities are one of the oldest internationals although perhaps they do not always publicize themselves as such.

When it was first proposed to the University Court last year that this particular Symposium should be held in St. Andrews, I know there was very great pleasure among all its members. We did appreciate it as a compliment not only to the University but, in particular, to the University Observatory and to its Director, Professor Stibbs. Although St. Andrews is a small University, perhaps because it is a small University, I think we have always been particularly proud of our special astronomical tradition. It is one of the things that sets us apart from other small Universities.

It is true, of course, that in its present form our Department of Astronomy is of relatively recent date. A separate Department only began in 1938; the modern Observatory building was only started in 1941, and the Chair itself was only created in 1950. But the tradition of Astronomy in St. Andrews goes back 300 years to the great figure of James Gregory, the contemporary and friend of Sir Isaac Newton. Had he been a member of a larger University, had he been the citizen of a larger country, I think perhaps James Gregory would be more widely known than he is. But by any reckoning, I think he was a remarkable man. He was the first incumbent of the Regius Chair of Mathematics founded in 1668 by Charles II. He had already in his *Optica Promota* in 1663 put forward suggestions for a two mirror combination for a reflecting telescope which in many respects anticipated, or at least preceded, alternative arrangements put forward five years later by Newton, and before similar proposals a few years after that by Cassegrain. Gregory built and equipped a University Observatory in St. Andrews which has now, I fear, disappeared, and we can only show you the site and some of the instruments that he used. But it is important, and I think certainly a matter of note for historians, that in this small town and small University, in the second half of the 17th century, Gregory in St. Andrews was founding an Observatory at the same time as Louis XIVth was founding his Observatory in Paris and before Charles II founded the English Observatory at Greenwich in 1675. And in Scotland, particularly in St. Andrews, we feel that Gregory was part of that great European intellectual renaissance, indeed, intellectual revolution particularly in the fields of Mathematics and Astronomy, that marked the second half of the 17th century. There are many marks of Gregory's life and career still to be found in the University. Some of his instruments, for example, are still preserved. If you go to Upper Parliament Hall in the University Library, you will see the shell, if not the lens, of one of Gregory's telescopes, the bracket that he used and the wall clock, now transformed into a long case clock, that he is believed to have used in making his observations and calculations. The name itself is preserved in our Second Chair of Mathematics, and it is also commemorated in the Cassegrain Schmidt Telescope which was built in St. Andrews after the war, very largely in the Observatory and which was, for a time, the largest telescope of its kind in Great Britain.

But, after that efflorescence in St. Andrews, the study of Astronomy undoubtedly declined in the 18th century. Politics and poverty are two great enemies of Universities, and St. Andrews suffered from both for some 150 years after Gregory left St. Andrews. The study of Astronomy did not revive in any real sense until the 19th century under the wing of the Chair of Mathematics, and the full recovery was reserved until our own time. But the tradition and the name of Gregory has never disappeared from the University, and no one or very few visitors who go over Parliament Hall can fail to be reminded of him. But, of course, tradition is only of value when it promotes and encourages the future, and in St. Andrews we look forward with confidence to the future of Astronomy in this University. Particularly, I think, we are looking forward to the next five years when the completion of the Anglo-Australian Telescope is clearly going to widen the opportunities for observational Astronomy and during which we hope that at least some of the equipment used on that telescope will originate from work in progress in St. Andrews.

And now I think I must not detain you any longer. You have your work to do, and I am only acting as a hindrance to it. All I want to add to what I have said is that we do wish you a very profitable Symposium. We also hope that you will enjoy your stay here. I know that arrangements have already been made for you to see other parts of Scotland but we hope that in your zeal to see the Highlands you will not forget to explore the more miniature attractions of St. Andrews, and that you will see as much as possible of the University while you are here.

TABLE OF CONTENTS

1. THE WHITE DWARFS

Discovery and Observation

W. J. LUYTEN

University of Minnesota, Minneapolis, Minn., U.S.A.

1. Introduction

White dwarfs have been known for slightly more than fifty years; with their planet-like diameters, and stellar-like masses their densities are enormously much higher than those of 'normal' stars. They thus appear to represent a state of matter totally unknown and possibly unattainable on earth.

The history of their original discovery and their early theoretical explanation has been told so often that we need not dwell on this now. Suffice it to say that while in the beginning they appeared to be rather rare in space, it is now generally believed that they constitute a substantial fraction of all known stars and represent the near-final stage of stellar evolution.

Using modern techniques, white dwarfs are fairly easy to discover, and several thousand are now known, but because of their low luminosity the follow-up observations to determine their astrophysically important properties are difficult and generally require very large telescopes.

2. Methods of Discovery

In the main there are two direct techniques for finding them, both based on the fact they are of low luminosity and generally of high temperature, therefore blue or white in color. The classical technique consists of finding first, faint stars with large proper motions, i.e. stars which, statistically, must be of low luminosity, and then subsequently determining the colors and thus select those objects that are whiter then the ordinary main-sequence stars. An entirely new method is due to Zwicky and Humason [1] who simply searched for faint stars that are blue in color, in regions of the sky where few, if any distant objects are expected, i.e. in front of obscuring clouds in the Milky Way, or in the direction of the galactic poles where the star-density is presumed to decrease rapidly with distance.

Applying the first of these methods in the Bruce Proper Motion Survey of the Southern Hemisphere, and, more recently, in the Palomar-Schmidt Proper Motion Survey, I have now found, and published more than three thousand white dwarfs – certain, probable, and possible. These will be discussed in more detail later.

The second method first led to the publication of the famous list of 48 Faint Blue Stars by Humason and Zwicky [1]. The fifteen stars found in the Hyades region did, indeed, turn out to be mainly white dwarfs, but among the thirty-three situated near

Luyten (ed.), White Dwarfs, 1–7. All Rights Reserved.
Copyright © 1971 by the IAU.

the North Galactic Pole, only two proved to be 'classical' white dwarfs, most of the others being the first representatives of a new class of objects, the 'Faint Blue Stars' in high galactic latitude.

Following up Zwicky's search, first at the Steward Observatory, later with plates taken at the Michigan, Dyer, Tonantzintla, and Palomar Observatories, I have published more than 20000 of such faint blue stars [2], 8700 of them jointly with Haro [3]. In order to obtain at least preliminary information on the number of white dwarfs among them I have determined proper motions for as may of these as old plate material could be found for, and have recently published a general catalogue [4] of 951 proper motions for such stars. In addition to our surveys similar surveys have been made, and lists of faint blue stars published by Iriarte and Chavira [5], Feige [6], Cowley [7], Rubin [8], Sanduleak and Philip [9], and Richter and Richter [10].

From an analysis of these proper motions, including preliminary determinations of the solar motion and secular parallaxes I found that among these faint blue stars the percentage of white dwarfs increases from virtually zero at $m = 13$ to about 10% at $m = 15$ or 16. For stars fainter than this limit the proper motion alone can hardly be considered as conclusive evidence: e.g. a white dwarf of the o_2 Eridani type with $M = +11$ will, at apparent magnitude 18, have a parallax of only $0''.004$ and hence, even near the galactic poles cannot be expected to have a proper motion much larger than $0''.040$ which is diffcult to determine with certainty from plate material that now exists.

As it was rapidly becoming clear that among very faint blue stars one could expect to find increasingly larger numbers of quasi-stellar objects, Sandage and I [11] embarked on a program of determining motions, photo-electric colors and spectra for as large a number of faint blue stars as we could observe in several representative regions in the sky. Since we used plates taken with the 48 inch Palomar-Schmidt telescope our limit was generally about $m = 19$; Sandage alone [11] continued the search to $m = 21$ on plates taken with the 200-inch telescope. From this work we obtain Table I giving the expected distribution of the various types of objects found among these faint blue stars.

TABLE I

	QSS and QSG	Main sequence	Horizontal branch	Subdwarfs	White dwarfs	Proper motion
13	–	15	25	60	–	$0''.6$
15	5	5	10	68	12	0.20
17	5	–	5	65	25	0.06
19	20	–	–	50	30	0.02
21	50	–	–	–	50	0.010
		160000	63000	20000	1000	

Note – The last column here gives the expected proper motion for a white dwarf of given magnitude; the last line gives the expected distance in parsecs of the various objects if of the 21st magnitude. It is possible that ultimately we should fit a column for U Geminorum variables, of expected absolute magnitudes between +6 and +9, between the columns for white dwarfs and subdwarfs, but too little is known at the present time about their frequency in space.

Spectroscopists [12] have usually claimed much larger percentages of white dwarfs among the brighter specimens of Faint Blue Stars, especially for the 13–16th magnitude. For a number of these I have subsequently determined proper motions, and generally found these to be so small as to indicate that if these objects are genuine white dwarfs of absolute magnitudes around +9 or fainter, they must all have very small tangential velocities which, at least near the galactic poles would be unexpected.

3. Colors and Spectra

The photometry of White Dwarfs, including the determination of accurate photo-electric colors will be adequately dealt with by Eggen, while the spectroscopic analysis will be covered by Greenstein and very little need be said here. Suffice it to point out that while the spectra of the first few white dwarfs found showed some similarity to those of main-sequence A stars it soon became apparent that the very high values of the surface gravity as well as possible Stark and magnetic effects would cause the spectroscopic features of most white dwarfs to be very different from those of ordinary stars.

It was for this reason that I proposed [13], in 1945, to use the letter D for the classification of white dwarf spectra, this to be followed by the usual, B, A, F, G, or K if the spectrum either somewhat resembled that of ordinary stars so classified or if the star had the same color as stars of those spectral classes. The further designation DC was reserved for stars showing featureless continuous spectra. To date, I believe that no degenerate object which would deserve the classification DM has been identified.

4. Parallaxes and Luminosities

At the present juncture parallaxes are perhaps the most urgently needed data for White Dwarfs. As of now trigonometric parallaxes have been published for fewer than twenty but this situation is on the point of being greatly improved now that the results from the U.S. Naval Observatory parallax program are becoming available. Further parallaxes may be derived for white dwarfs which are components of binaries in which the other component appears to be an ordinary main-sequence star and hence a reliable spectro-photometric parallax can be derived. Finally a number of white dwarfs have been identified as belonging – with high probability – to galactic clusters for which the distance is known: 20 in the Hyades, 4 in Praesepe, and 1 each in the Pleiades and M 67.

When the usual H-R diagram is made up for these stars it is seen that degenerate stars occupy a broad area to the lower left of the diagram, running roughly parallel to the main sequence from about $M_V = +8.5$ for very blue stars to at least $M_V = +17$ for degenerate stars with color indices of $+1.0$ or more.

For the more than three thousand possible white dwarfs found in the proper motion surveys it is not yet possible to make up such a diagram and all we can do is to plot the rough colors of the stars against the reduced proper motions H, where $H = m + 5 + 5$

$\log\mu$ or, also $H = M + 5\log T$. Because of this, the full uncertainty in our knowledge concerning the kinematics of white dwarfs would come into play in such a diagram. As the guiding principle in selecting stars for the white dwarf catalogue I used the criterion that any star which, if classified as a main-sequence object, would have a tangential velocity of more than 500 km/sec, appeared likely to be degenerate. For the stars classified as having colors b, a, or f this leads to little trouble, and probably the vast majority of the 1600 stars so listed in the catalogue are genuine white dwarfs. On the other hand, the attrition, or perhaps I should say the casualties, among stars yellower than this can be expected to increase, and become quite large for those listed as of color k, on the one hand because these crude colors, determined from the Palomar Survey plates may often give a value of '*k*' for a star of actual spectral class M, and on the other hand because among these stars will be found many subdwarfs with exceedingly high velocities, and ultraviolet excesses. However I know of no other way at the present time, in which yellow degenerate stars can be identified.

Whether DM stars exist we do not yet know – the reddest definitely known degenerate star having a color index of + 1.1 Among the more than 150 double stars with one white dwarf component I have found [14] that invariably the main-sequence component is bolometrically the more luminous. Among the rather few double stars where both components appear degenerate, the fainter component is invariably the yellower one. There remain five double stars where the brighter component appears to be definitely degenerate and the fainter one has the color m – I suggest that these hold out the best prospect of being M-type degenerates although, of course, it is always possible that they are extreme M-type subdwarfs with ultraviolet excess.

To forestall any possible criticism that – if a large majority of the g and k probable white dwarfs I have listed prove to be only very high velocity subdwarfs – I have led the spectroscopists astray and caused them much unnecessary work I should like to point out that for a number of years the spectroscopists have urgently asked for lists of candidates for yellow-degenerates. In the course of the Bruce and Palomar Proper Motion surveys I have probably looked at 100 million stars and from them have made up this list containing some 3000 white dwarfs, including 1484 possible g and k degenerates, representing a screening, or refining, of better than 60000 to 1. Even if only 5% of these yellow degenerate candidates were to prove genuine I would still say it has been well worth it – and the spectroscopists should not complain. And, at any rate, the by-product, i.e. the yellow subdwarfs with extraordinary high velocities are not exactly wasted either.

It now seems generally accepted that neutron stars, first postulated by Zwicky and Baade [15], exist – these are assumed to have densities of the order to 10^{14} and up. The typical white dwarf has a density of the order of, perhaps, 10^5. Whether stars intermediate to these, the 'pygmies' as also first postulated by Zwicky – with densities of perhaps 10^7 or 10^8, also exist is not yet known. The few apparently white or blue stars with exceedingly large proper motions which would be candidates for this are claimed by the spectroscopists to be somewhat yellower than first thought, and ordinary degenerates of very high space motion. The only two parallaxes available for such stars – LP 9-231 and LP 768-500 – certainly agree better with the spectroscopists point of view than with the pygmy supposition.

On the other hand, if the spectroscopists were correct in their claim that there exist large numbers of Faint Blue Stars which are in reality ordinary white dwarfs but with very small tangential velocities, and further that all apparent pygmies are ordinary white dwarfs, with excessively high tangential velocites, and, finally, that most apparent yellow degenerates are really subdwarfs with similarly high velocities, then we shall be faced with a rather peculiar distribution of velocities. Not to mention the fact that many of these stars should not only be escaping from the galaxy, but, unless they are younger than 10^8 yr they should have escaped long ago. Again, the answer seems to be: what is needed is more parallaxes of very faint stars, especially of the 19th and 20th magnitude.

5. Masses and Redshifts

The first two white dwarfs discovered, Sirius B and o_2 Eridani B were both components of binaries, and their masses were known beforehand. Since that time, the companion to Procyon, also with a known mass, has been identified as a probable white dwarf. However, the companion to Sirius has often been suspected of being a close double, and its mass is rather larger than expected, while the companion to Procyon is so faint and so close to its primary that it is virtually impossible to obtain reliable values for its apparent magnitude, color, and spectrum. This leaves o_2 Eridani B as the only white dwarf for which the mass, luminosity, color, and spectrum, hence also the surface temperature are known with reasonable accuracy – surely not an auspicious base on which to erect a whole theory of white dwarfs.

I have repeatedly pointed out that accurate astrometric observations with large telescopes made on binaries containing white dwarf components could give us, in a reasonable time, at least an indication of orbital motions, from which, in turn, statistical estimates of the masses might be made, but until now no such observations appear to have been made. Using a large, and rather heterogeneous mixture of photographic plates, I attempted, in 1961 [17], to determine such preliminary orbital motions for some 17 binaries with white dwarf components, and by comparing these with similar orbital motions derived for a number of control binaries composed of apparently normal main-sequence stars and possessing very similar apparent magnitudes, separations, and proper motions, and to derive a statistical indication of the ratio of the masses of typical white dwarfs to those of main-sequence stars of the same luminosity. While the results are far from conclusive, they do indicate that the masses of 'classical' white dwarfs are of the order of 1.5 times the masses of similar main-sequence stars.

Now that nearly two hundred binaries containing white dwarf components are known – including more than a dozen pairs where both components appear to be degenerate – I urge again that those who have access to large telescopes begin immediately to take the first epoch plates necessary for the ultimate determination of orbital motions and of statistical masses. I believe that at least at present there is no other way in which reliable values for the masses of degenerate stars can be determined.

Theory demands a close relationship between the mass, radius, and spectral red shift of a white dwarf but it now appears that this relation is not quite as simple as we used to believe. Moreover, real red shifts cannot be determined for single white dwarfs (except, again, statistically, and this would involve some assumptions as to the kinematics involved) while, in addition, many of the most interesting white dwarfs have almost featureless spectra in which red shifts cannot be reliably determined. Hence, in final analysis, we fall back again on astrometric observations and Kepler's laws for the reliable determination of masses for degenerate stars.

6. White Dwarfs in Clusters

Present theory suggests there should be more white dwarfs in old clusters such as M 67, the Hyades and Praesepe than in young clusters such as the Pleiades and h and χ Persei. To test this I have blinked pairs of blue and red Palomar Survey plates [18] for a number of galactic clusters nearer than 800 parsecs (in which accordingly, a classical o_2 Eridani B type of white dwarf would appear brighter than 20.7) but found no indication of such stars in M 36, M 38, M 39, NGC 129, and NGC 6885. Possibly a few may exist in IC 4665, M 34, and NGC 752 but repeated searches [19] have as yet shown no degenerate stars belonging to the Ursa Major Cluster (except possibly the companion to Sirius). M 67 is so distant that while several possible white dwarfs have been indicated, only one has been identified with reasonable certainty to be a member. Similarly only one probable white dwarf belonging to the Pleiades has been announced [20] while four have been identified with reasonable certainty in Praesepe [21]. Only in the Hyades do we know a substantial number of white dwarfs which are cluster members. The first two of these were found by Van Rhijn and Raimond [22], the next several by Humason and Zwicky [23] some more by myself [24] and by Van Altena [25], while I have [24] also indicated a few yellow degenerates as possible cluster members. All in all we may now have some twenty such stars in the Hyades.

7. Frequency in Space

From my proper motion surveys I have estimated [26] that white dwarfs constitute only a few percent – 2.3% ± 0.3% of all stars in space while theoreticians have generally derived much higher values – up to 10%. The difference may be due largely to a matter of definition – my own values applying only to really blue or white degenerate stars whereas the theoretical estimates include yellow degenerates as well. Further, if indeed large numbers of Faint Blue Stars should prove to be genuine white dwarfs with very small tangential velocities my estimates will have to be raised considerably. On the other hand, if most of the proper motion stars now designated as possible white dwarfs should indeed turn out to be high-velocity subdwarfs then the theoretical estimate would have to be substantially reduced. When the Palomar-Schmidt proper motion survey has been completed and when several hundred parallaxes for white dwarfs down to the seventeenth and eighteenth magnitude have been determined we shall be in a better position to arrive at a really reliable estimate.

Meanwhile the rough figure of 5% may represent an acceptable compromise.

References

[1] Humason, M. L. and Zwicky, F.: 1947, *Astrophys. J.* **105**, 85.
[2] Luyten, W. J.: 1933–1968, *A Search for Faint Blue Stars*, Nos. I–L.
[3] Haro, G. and Luyten, W. J.: 1962, *Bol. Obs. Tac. y Ton.*, No. 22.
[4] Luyten, W. J.: 1969, *A Search for Faint Blue Stars*, No. L.
[5] Chavira, E. and Iriarte, B.: 1957, *Bol. Obs. Tac. y Ton.*, No. 16; also No. 18, 1959.
[6] Feige, J.: 1958, *Astrophys. J.* **128**, 267.
[7] Cowley, C. R.: 1959, *Astron. J.* **63**, 484.
[8] Rubin, V.: 1967, *Astron. J.* **72**, 59.
[9] Sanduleak, H. and Philip, A. D.: 1958, *Bol. Obs. Tac. y Ton.* **4**, 253.
[10] Richter, N. and Richter, L.: 1965, *Mitt. Karl Schwarzschild Obs.*, No. 24.
[11] Sandage, A. R.: 1967, *Astrophys. J.* **148**, 767; also **155**, 913.
[12] Kinman, T. D.: 1965, *Astrophys. J.* **142**, 1241; also Greenstein, J. L.: 1966, *Astrophys. J.* **144**, 496.
[13] Luyten, W. J.: 1945, *Astrophys. J.* **101**, 131.
[14] Luyten, W. J.: 1969, *Proper Motion Survey with the 48 inch Schmidt Telescope*, No. XVIII.
[15] Baade, W. and Zwicky, F.: 1934, *Proc. Nat. Acad. Sci. U.S.* **20**, 254.
[16] Luyten, W. J.: 1956, *A Search for Faint Blue Stars*, No. VIII; also 1969, *Proper Motion Survey with the 48 inch Schmidt Telescope*, No. XVIII.
[17] Luyten, W. J.: 1961, *Publ. Astron. Observ. Univ. Minn.* **3**, No. 9.
[18] Luyten, W. J.: 1958–1963, *A Search for Faint Blue Stars*, Nos. X–XXXII.
[19] Luyten, W. J.: 1958, *A Search for Faint Blue Stars*, No. XIV.
[20] Luyten, W. J. and Herbig, G.: 1960, *Harv. Ann. Card*, No. 1474.
[21] Luyten, W. J.: 1962, *A Search for Faint Blue Stars*, No. XXXI.
[22] Van Rhijn, P. and Raimond, J. J.: 1934, *Monthly Notices Roy. Astron. Soc.* **94**, 508.
[23] Humason, M. L. and Zwicky, F.: 1947, *Astrophys. J.* **105**, 85; also *A Search for Faint Blue Stars*, No. I, 1952.
[24] Luyten, W. J.: 1969, *Proper Motion Survey with the 48 inch Schmidt Telescope*, No. XVII.
[25] Van Altena, W. F.: 1966, *Astron. J.* **71**, 482.
[26] Luyten, W. J.: 1958, foll. *A Search for Faint Blue Stars*, No. XVI.

2. RED SUBLUMINOUS STARS

O. J. EGGEN

Mount Stromlo and Siding Spring Observatories,
Research School of Physical Sciences,
The Australian National University, Camberra, Australia

(UBVRI) observations of known subluminous stars were published in a previous discussion (Eggen, 1970a) where the bulk of the objects, both red (RSL) and blue (BSL) populated a sequence extending for M (I) near $+11^m5$ at $(R-I)$ near 0^m0 to M (I) of $+15^m$ near $R-I = +0^m3$. However, four RSL stars appeared to form a separate sequence near $M(I) = +11^m$ and $R-I$ between $+0^m3$ and $+0^m7$. These four stars, plus an additional object (LTT 2236) are listed in Table I together with the source of luminosity; the weights of the trigonometric parallaxes are given in parentheses following the parallax. These five stars are represented in Figure 1 by crosses; the top of the steep sequence of subluminous stars is shown as the hatched region in the figure. For reference, Figure 1 also contains (1) the Hyades main sequence (Eggen, 1969b), indicated by a broken curve, (2) the evolved main sequence of the old disk population (Eggen, 1970b), shown by open circles, and (3) the subdwarfs, which are listed in

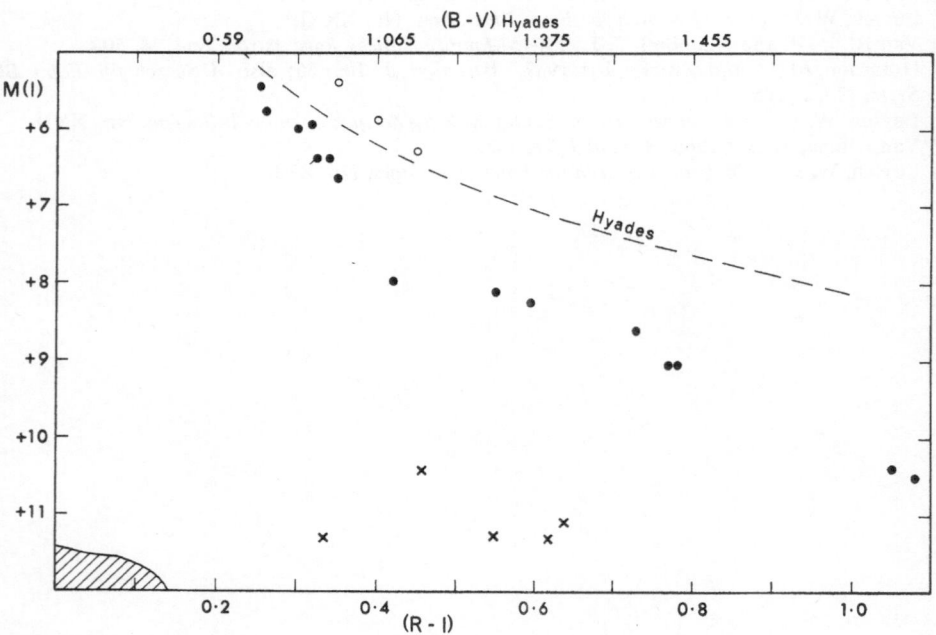

Fig. 1.   Evolved, old disk population main sequence stars (open circles), Hyades cluster stars (broken curve), subdwarfs (filled circles, and RSL stars (crosses) in the $(M_I, R-I)$ plane. The top of the steep sequence of subluminous stars (white dwarfs) is shown as the hatched area in the lower left hand corner.

TABLE I

Subluminous red stars

Name	EG	V_E	B − V	U − B	R	R − I	M(I)	Remarks
G 14-24	96	$12^m.78$	$+0^m.75$	$0^m.00$	$12^m.32$	$+0^m.335$	$+11^m.3$	0".074(8), Ross 974
LTT 2236	–	13.19	+1.16	+0.72	12.72	+0.46	+10.4	0".042(8)
GH 7-138	32	15.65	+1.05	+0.32	14.81	+0.55	+11.3	Hyades cluster
Wolf 1037	–	14.19	+1.42	+1.22	13.24	+0.62	+11.3	0".054(8), G 18-51
HR 58973B	–	13.50	+0.86	+0.15	12.72	+0.64	+11.1	βTrA cpm, LTT 6333

O. J. EGGEN

TABLE II
Late type subdwarfs

Yale	V_E	B − V	U − B	R	R − I	M(I)	π_{tr}(wt)
887.0	8.51	+ 0.86	+ 0.37	8.10	+ 0.34	+ 6.4	0.053 (58)
948.1	11.85	+ 1.37	+ 1.02	10.90	+ 0.37	+ 8.6	0.048 (12)
1181.0	8.90	+ 1.53	+ 1.07	7.84	+ 0.77	+ 9.05	0.251 (18)
1857.0	8.32	+ 0.62	− 0.14	8.15	+ 0.26	+ 5.8	0.038 (61)
2392.0	8.07	+ 0.595	− 0.025	7.68	+ 0.255	+ 5.45	0.040 (16)
2512.0	11.02	+ 1.40	+ 0.95	10.04	+ 0.775	+ 9.05	0.091 (7)
2745.0	6.45	+ 0.75	+ 0.17	6.08	+ 0.30	+ 6.0	0.110 (36)
3044.0	10.85	+ 1.00	+ 0.68	10.23	+ 0.42	+ 8.0	0.043 (15)
3252.0	13.43	+ 1.58	+ 1.08	12.21	+ 1.05	+ 10.4	0.070 (12)
3425.0A	9.10	+ 0.78	+ 0.13	8.71	+ 0.32	+ 6.4	0.040 (35)
3425.0B	9.45	+ 0.85	+ 0.30	9.00	+ 0.35	+ 6.65	
3669.0	7.52	+ 0.84	+ 0.24	7.20	+ 0.32	+ 5.9	0.063 (40)
3783.1	12.73	+ 1.64	+ 1.20	11.63	+ 1.08	+ 10.5	0.099 (10)
5741.1A*	12.13	+ 1.28	+ 1.05	11.26	+ 0.55	+ 8.1	0.030 (*)
5741.1B*	12.84	+ 1.39	+ 1.16	11.39	+ 0.595	+ 8.2	

* The values for the bright and faint components are, respectively, 0″.024 (14) and 0″.042 (7).

Table II. In addition to the photometric results, Table II contains the Yale Parallax Catalogue number, the mean trigonometric parallax and its weight, in parentheses, as well as the resulting values of M(I).

The luminosity of any one star in Table I cannot be accepted without some reserve.

(1) The single trigonometric parallax, determined at the Cape, for G 14-24 is large and apparently (Greenstein and Eggen, 1966) well determined. The total annual proper motion is 0″.47, leading to a tangential velocity of 275 km/sec if the star were a subdwarf ($\pi = 0″.008$). The radial velocity is + 156 km/sec so the resulting total space motion of 320 km/sec does not eliminate the possibility that this is a subdwarf.

However, the requirement that the trigonometric parallax is in error by a factor of 10 is very unlikely. The spectrum of the star is indistinguishable from that of a weak-lined subdwarf (Eggen and Greenstein, 1965).

(2) The single trigonometric parallax of 0″.042 (wt. 8) for LTT 2236 has been determined at the Cape Observatory (Yale, No. 1216.1). If the star is a subdwarf the parallax is 0″.012 and the total annual proper motion of 0″.52 would give a tangential velocity of 200 km/sec. Additional parallax determinations are needed.

(3) The membership of GH 7-138 (van Altena, No. 71) in the Hyades cluster seems well established on the basis of three determinations of the proper motion (van Altena, 1969; Table IIIc).

(4) The single trigonometric parallax determination for Wolf 1037 (LTT 16591) is from the Mount Wilson Observatory. As a subdwarf the parallax would be 0″.015 and the large annual proper motion of 1″.68 would lead to a tangential velocity of 530 km/sec. The radial velocity of − 160 km/sec, which may be variable, gives a total space motion of 550 km/sec. As for G 14-24, the spectrum is indistinguishable from

that of a very weak-lined subdwarf, K-type star (Joy, 1947). Additional parallax observations are needed.

(5) Spectra of LTT 6333 are not available. The luminosity is based on the assumption that the proper motion of LTT 6333 (0".40 223°) is common with that of HR 5897 (β Tr A, 0".42 205°) which is 157" distant. The single trigonometric parallax determination for HR 5897 is 0".078 (wt. 7) from the Cape Observatory and the photometric value, from (u, b, v, y) photometry is 0".063.

The five late type, subluminous stars discussed above also show abnormal values of the ultraviolet excess. This is demonstrated in Figure 2(a) where the continuous

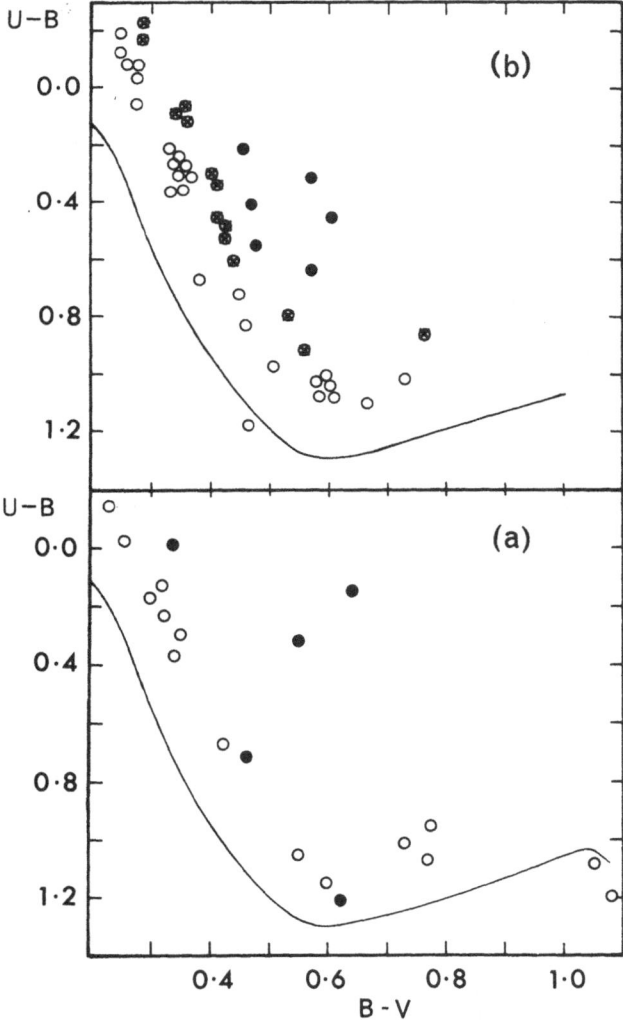

Fig. 2. The continuous curve in both panels represents the main sequence stars of the Hyades cluster. The open and filled circles in (a) represent the subdwarfs and subluminous stars in Tables II and I, respectively. The open circles in (b) represent probable subdwarfs (sd?) in Table III, the crossed circles, possible RSL stars and the filled circles, probable RSL stars.

TABLE III
Possible RSL stars

Star	V_E / R	$B-V$ / R-I	$U-B$ / N, N	μ / θ	Sp. / $\delta(U-B)$	$\Delta(U-B)$ / $\Delta(B-V)$	Type	π(sd) / T(sd)
L 233-10	14.46	+0.835	+0.22	0″.21	g	+0.235	RSL	0″.007
LTT 929	14.00	+0.465	3,2	173°	+0.29	+0.90		150
L 729-4	13.26	+0.985	+0.40	0.46	k	+0.095	RSL?	0.010
LTT 1244	12.82	+0.41	2,2	170	+0.40	+0.575		200
L 127-50	15.20	+0.95	+0.27	0.24	k	+0.04	sd?	0.003
LTT 1346	14.80	+0.36	2,3	178	+0.47	+0.545		400
L 298-26	13.73	+0.91	+0.30	0.38	g	+0.09	sd?	0.006
LTT 1358	13.30	+0.365	2,2	90	0.36	+0.53		300
LTT 1561*	14.49	+0.92	+0.25	0.31	k	+0.05	sd?	0.004
G77-43	14.11	+0.345	3,3	149°	+0.43	+0.505		350
L 54-9	13.64	+0.47	−0.22	0.24	g	+0.34	RSL?	0.0035
LTT 1607	13.40	+0.28	3,1	84	+0.23	+0.635		300
LTT 1721*	14.12	+0.825	+0.07	0.27	k	+0.155	RSL?	0.0045
G 160-6	13.78	+0.355	2,2	190	+0.415	+0.73		300
Ross 580*	13.04	+0.79	+0.08	0.23	k-m	+0.18	RSL?	0.0075
G 160-8	12.58	+0.35	2,2	150	+0.33	+0.70		150
L 807-30	14.23	+1.26	+0.92	0.44	g	+0.09	RSL?	0.011
LTT 2030	13.53	+0.56	2,3	155	−	+0.36		200
L 665-68	12.95	+0.67	−0.08	0.40	k	+0.105	sd?	0.004
LTT 2175	12.75	+0.265	3,3	138	+0.285	+0.43		450
Strand 11*	13.28	+1.015	+0.48	0″.32	−	+0.08	RSL?	0″.010
G 99-8	12.78	+0.42	2,2	72	+0.39	+0.52		150
Strand 12	14.83	+1.20	+1.19	0.34	−	−0.01	?	
G 99-9	13.80	+0.475	1,2	77	−	−0.05		
G 99-47	14.08	+0.61	−0.11	1.07	−	+0.165	sd?	0.0025
−	13.83	+0.25	4,3	208	+0.25	+0.39		2000
L 595-22	12.25	+0.38	−0.25	0.32	−	+0.09	*	*
LTT 2415	12.20	+0.155	5,2	124	+0.27	+0.265		
L 812-11	13.14	+1.00	+0.53	0.88	g	+0.095	RSL?	0.011
LTT 2535	12.58	+0.42	4,4	140	+0.31	+0.47		350
L 455-129	13.66	+1.34	+1.04	0.86	f:	+0.035	sd?	0.012
LTT 2826	13.62	+0.605	2,2	154	−	+0.25		450
G 112-28	13.72	+1.12	+0.82	1.01	−	+0.05	sd?	0.011
−	13.10	+0.465	3,2	169	−	+0.295		450
G 113-36	15.45	+1.14	+0.64	0.30	−	+0.215	RSL	0.0065
−	14.73	+0.57	1,3	141	−	+0.645		200
LTT 3144*	11.97	+0.93	+0.51	1.00	k-m	+0.155	RSL?	0.018*
G 113-40	11.41	+0.41	3,3	157	+0.19	+0.465		250
L 190-19	13.30	+1.00	+0.675	0.39	g	+0.025	sd?	0.0085
LTT 3807	12.73	+0.38	2,3	272	+0.165	+0.205		200
L 36-61	13.08	+1.12	+0.79	0.43	k	+0.18	RSL?	0.0185
LTT 3862	12.22	+0.53	1,1	121	−	+0.46		100
L 260-53	13.27	+1.47	+1.10	1″.11	g	−0.08	sd?	0″.024
LTT 5622	12.55	+0.67	1,2	249	−	+0.19		200
L 477-3	11.90	+0.88	+0.28	0.52	g:	+0.06	sd?	0.011
LTT 5864	11.62	+0.335	2,2	245	+0.32	+0.43		200
L 478–87	13.24	+0.925	+0.365	0.26	a-f	+0.055	sd?	0.007
LTT 5889	12.88	+0.355	2,2	270	+0.335	+0.43		200

Table III (continued)

Star	V_E / R	B−V / R−I	U−B / N, N	μ / θ	Sp. / δ(U−B)	Δ(U−B) / Δ(B−V)	Type	π(sd) / T(sd)
L 262-45	13.80	+0.92	+0.31	0.32	f	+0.145	RSL?	0.007
LTT 5944	13.34	+0.40	2,1	213	+0.37	+0.64		200
Ross 1038*	12.33	+0.765	+0.055	0.96	m	+0.025	sd?	0.0065
G 15-13	12.05	+0.275	2,1	207	+0.31	+0.355		700
L 624-39	12.20	+0.845	+0.105	0.39	g	+0.115	sd?	0.010
LTT 6307	11.91	+0.345	2,3	186	+0.425	+0.65		200
L 628-48	12.90	+0.68	−0.16	0.30	f	+0.11	RSL?	0.0045
LTT 6447	12.78	+0.275	3,2	231	+0.385	+0.57		300
−3°3968A*	9.63	+0.74	+0.09	0.75	G5	+0.17	RSL?	0.031*
G 17-25	9.24	+0.32	4,6	193	+0.23	+0.555		100
−3°3968B	13.88	+1.43	+1.46:	0.73	−	−0.05	?	
G 17-27	13.01	+0.63	2,4	190	−	−0.18		
G 17-28	14.28	+1.42	+1.07	1.26	−	−0.045	sd?	0.012
−	13.67	+0.605	2,3	226	−	+0.22		500
G 19-17	14.67	+1.42	+0.46	0.28	−	−0.045	RSL	0.011
−	13.86	+0.605	1,3	189	−	+0.83		100
L 754-45*	11.52	+0.535	−0.19	0″.24	g	+0.195	sd?	0″.0085
LTT 3966	11.13	+0.255	3,2	253	+0.255	+0.495		150
L 754-46	13.10	+0.84	+0.23	0.24	g	+0.13	sd?	0.0085
LTT 3967	12.52	+0.35	5,2	253	+0.29	+0.55		150
G 163-59	14.86	+1.46	+0.86	1.14	−	−0.035	RSL?	0.016
−	13.84	+0.77	1,2	202	−	+0.44		350
L 611-42	12.78	+0.56	−0.18	0.31	g:	−		
LTT 4210	−	−	4,−	287	+0.27	−		
L 611-43	14.94	+1.00	+0.54	0.31	g:	+0.355	RSL	0.009
LTT 4211*	14.05	+0.57	5,1	287	+0.30	+0.745		150
L 614-137	14.30	+1.36	+1.02	0.60	k	+0.005	sd?	0.0115
LTT 4667	13.52	+0.585	2,2	154	−	+0.265		250
L 105-2	13.10	+0.695	−0.04	0.24	a	+0.115	sd?	0.005
LTT 4896	12.80	+0.28	2,2	300	+0.295	+0.475		250
L 328-123	14.25	+1.08	+0.62	0.45	k	+0.04	RSL?	0.0065
LTT 4953	13.84	+0.435	3,2	296	+0.35	+0.42		350
LTT 5074*	12.94	+0.95	+0.30	0.58	g	+0.02	sd?	0.0085
G 14-39	12.40	+0.35	3,2	288	+0.44	+0.48		350
L 196-36	12.28	+1.36	+1.08	0.30	g	+0.01	sd?	0.031
LTT 5220	11.49	+0.59	2,2	165	−	+0.31		50
L 547-141	12.69	+0.87	+0.23	0.47	k	+0.07	sd?	0.008
LTT 5472	12.25	+0.335	3,3	270	+0.35	+0.48		250
L 836-104	13.65	+0.73	−0.07	0.25	k	+0.03	sd?	0.0035
LTT 5560	13.42	+0.265	2,2	201	+0.375	+0.43		350
Ross 858	13.64	+1.31	+1.04	0″.61	k	+0.06	sd?	0″.018
LTT 6979	12.82	+0.595	2,3	210	−	+0.25		150
LTT 7424*	12.86	+1.00	+0.56	0.31	g	+0.19	RSL	0.0145
G 155-35	12.42	+0.475	2,2	214	+0.28	+0.58		100
L 1143-61*	13.76	+0.87	+0.37	0.59	k	+0.07	sd?	0.005
G 25-1	13.37	+0.335	2,2	197	+0.21	+0.34		600
Ross 770*	11.75	+1.19	+0.97	1.10	K4	+0.07	sd?	0.031
LTT 8417	11.00	+0.51	5,4	192	−	+0.245		150
L 716-108	13.76	+1.32	+1.00	1.05	m	+0.05	sd?	0.016
LTT 8975	12.87	+0.59	1,3	156	−	+0.29		250

Table III (continued)

Star	V_E	B − V	U − B	μ	Sp.	Δ(U − B)	Type	π(sd)
	R	R − I	N, N	θ	δ(U − B)	Δ(B − V)		T(sd)
LTT 9372*	13.57	+ 1.00	+ 0.42	0.40	k	+ 0.17	RSL	0.0155
G 157-20	13.16	+ 0.465	3,2	115	+ 0.42	+ 0.695		150
L 793-57	13.52	+ 0.88	+ 0.34	0.75	g	+ 0.07	sd?	0.0055
LTT 9765	13.05	+ 0.34	7,4	169	+ 0.26	+ 0.39		650

*LTT 1561:   Lowell, $\mu = 0''.31$, $\theta = 139°$.
LTT 1721:   Lowell, $\mu = 0''.31$, $158°$.
Ross 580:   LTT 1728. Lowell, $\mu = 0''.28$, $\theta = 149°$.
Strand 11/12:   Lowell, $\mu = 0''.29$ and $0''.29$, $\theta = 78°$ and $78°$.
LTT 2415:   Almost certainly a subluminous star.
LTT 3144:   Lowell, $\mu = 0''.96$, $\theta = 156°$. Trigonometric parallax (Y2019.0) is $0''.045$, $wt = 15$.
LTT 3966/7:   LDS 315, $27''300°$.
LTT 4210/1:   LDS 350, $40''256°$.
LTT 5074:   Lowell, $\mu = 0''.59$, $\theta = 290°$.
Ross 1038:   Lowell, $\mu = 0''.87$, $\theta = 213°$.
− 3°3968A: Also LTT 6621, $\mu = 0''.82$, $\theta = 191°$. The two stars are separated by 20′. The trigo-
            nometric parallax (Y3767.0) is $0''.046$, $wt = 17$.
LTT 7424:   Lowell, $\mu = 0''.30$, $\theta = 225°$.
L 1143-61:   Lowell, $\mu = 0''.66$, $\theta = 198°$. Also G 141-35, $\mu = 0''.61$, $\theta = 199°$.
Ross 770:   The trigonometric parallax (Y5100.0) is $0''.008$, $wt = 6$.
LTT 9372:   Lowell, $\mu = 0''.49$, $\theta = 116°$.

curve represents the Hyades cluster stars and the open circles represent the subdwarfs in Table II. The RSL stars in Table I are shown in Figure 2a as filled circles and the resulting values of Δ(U − B) = + $0^m.8$, + $1^m.0$ and + $1^m.1$ for G 12-24, GH 7-138 and HR 5897B, respectively, are considerably larger than expected from the abundance effect alone.

The high ultraviolet (HUV) excess criterion for isolating possible RSL stars was previously applied to some 1000 southern proper motion stars (Eggen, 1969a), About 100 objects, or 10 percent of the proper motion stars, were found to be probably subluminous by this criterion. Observations of R and R − I are now available for 51 of these stars and are listed in Table III together with (1) the discoverer's number and the identifications in the LTT catalogue (Luyten, 1957) or lists published by Giclas and his associates (Lowell Observatory Bulletins); (2) the (UBV) photometry and the number (N, N) of (UBV) and (R, I) observations, respectively; (3) the annual proper motion and its direction; (4) the spectral type or color class (Luyten, 1957) and the ultraviolet excess, δ(U − B); (5) the values of Δ(U − B) and Δ(B − V) obtained from the (R − I, B − V) and (R − I, U − B) relations for Hyades cluster stars (Eggen, 1970b); (6) the probable classification of the star, discussed below, and (7) the photometric parallax and resulting tangential velocity if the star is assumed to a subdwarf and falling on the mean subdwarf sequence in Figure 1.

The stars in Table III are shown in the $(R - I, U - B)$ plane of Figure 2b where the continuous curve represents the relation for Hyades cluster, main sequence stars. The open circles in Figure 2b represent objects that, by analogy with Figure 2a, could be subdwarfs, the crossed circles are possible subluminous stars on the basis of the large ultraviolet excesses and the filled circles represent HUV objects that are probably subluminous. The filled circles are referred to as RSL objects in the penultimate column of Table III, the crossed circles as RSL? stars and the open circles as sd? stars. From a comparison with Figure 2a it is assumed that the sd? stars may include a few RSL objects, the crossed circles may include a few subdwarfs and the filled circles may all represent RSL stars.

The distribution of tangential velocities listed in the last column of Table III and based on the assumption that all of the stars are subdwarfs, adds some weight to the classification of six objects (filled circles in Figure 2b) as RSL stars. Omitting one obvious subluminous star, G 99-47, for which the subdwarf assumption leads to the improbable tangential velocity of 2000 km/sec, the mean value for the 26 'sd?' stars, or common proper motions systems, is $T = 300$ km/sec compared with 140 km/sec as the mean for the 6 RSL stars. That is, the stars assigned to the RSL classification on the basis of the ultraviolet excess almost certainly have smaller tangential velocities than the probable subdwarfs. The known RSL stars in Table I have a mean tangential velocity of 40 km/sec and the six stars shown as filled circles in Figure 2b also give a mean of 40 km/sec if the same mean luminosity, $M(I) = +11\overset{m}{.}1$, is assumed. The 'RSL?' stars, crossed circles in Figure 1b, give an intermediate value of 230 km/sec as the mean tangential velocity on the subdwarf assumption. Some of these stars are undoubtedly RSL stars and from a consideration of both the tangential velocity and the color, the most likely candidates in this group are Ross 580, G 99-8/9, LTT 3862 and $-3°3968$A, B. The trigonometric parallax listed in the notes to Table III for this last pair places the fainter component among the RSL stars, with $M(I) = +10\overset{m}{.}7$ at $R - I = +0\overset{m}{.}63$, but gives the brighter component values of $M(I) = +7\overset{m}{.}2$ at $R - I = +0\overset{m}{.}32$. Spectroscopic observations would be of great interest.

TABLE IV

Photometric parallaxes for probable RSL stars

Name	M(I) =		Name	M(I) =	
	+ 11.1	+ 14.5		+ 11.1	+ 14.5
LTT 929	0″.031	–	LTT 5220	0″.110	–
Ross 580	0.060	0.280	Ross 1038	0.072	0.350
G 99-8	0.055	–	− 3°3968	See text	–
G 99-47	0.031	0.150	LTT 7424	0.070	–
LTT 2415	0.060	0.150*	LTT 9372	0.048	–
LTT 3862	0.075	–	LTT 9765	0.048	0.230
LTT 4211	0.033	–			

* $M(I) = +13.0$

The most likely RSL stars in Table II are listed in Table IV together with the photometric parallax based on the assumption that, like the stars in Table I, the luminosities are near $M(I) = +11^m1$. The stars in Table IV also include the three 'sd?' objects in Table III with tangential velocities greater than 600 km/sec from the sub-dwarf assumption (G 99-47, Ross 1038 and LTT 9765). As RSL stars the five objects in Table IV with $(R-I)$ less than $+0^m35$ may populate the steeper sequence of sub-luminous stars (white dwarfs) for which the luminosities would be near $M(I) = +14^m5$ and in these cases the resulting parallax is also listed in the table.

All of the stars in Tables I and III are of interest but astrometric and spectroscopic observations of those in Tables I and IV are especially important and will probably settle the questions of the reality of the upper sequence of subluminous stars apparently defined by the objects in Table I.

In the previous discussion (Eggen, 1969a) attention was called to two common proper motion systems in which one component may be an RSL object. The $(R-I)$ photometry for these stars gives the following;

Star	V_E	$B-V$	$U-B$	R	$R-I$	N,N μ		θ	
G 22-9	10^m10	$+0^m705$	$+0^m15$	9^m84	$+0^m26$	2,3	0″305	192°	LTT 7511
-8	13.52	+0.96	+0.33	12.97	+0.41	4,3	0.305	192°	LTT 7512
LTT 5430	9.34	+0.815	+0.35	9.06	+0.295	4,2	0.22	170	
LTT 5428	13.98	+0.70	+0.20	13.58	+0.23	3,3	0.23	268	

The components of the first pair are separated by 33″ and those of the second by 300″. From the (R, I) photometry it appears that the first pair may consist of halo population subdwarfs with a parallax of 0″009 and tangential velocity of 160 km/sec. If this interpretation is correct, the radial velocity may be very large. If LTT 5430 is a normal, main sequence star the companion LTT 5428, has $M(I)$ near $+10^m2$ and is an RSL object with tangential velocity of 45 km/sec.

An additional common proper motion pair that deserves astrometric and spectro-scopic attention is the following:

Star	V_E	$B-V$	$U-B$	R	$R-I$	μ	θ
G85-44	12^m48	$+1^m50$	$+1^m00$	11^m28	+1.04	0″61	171°
G85-40	14.76	+1.15	+0.67	14.10	+0.46	0.61	170°

The components are separated by 1°2. The fainter component is also G83-53 (0″63, 172°), and G 97-16 (0″58, 168°) whereas the brighter component is also G 97-12 (0″61, 167°) and Ross 388 (0″59, 173°). In spite of the large separation, the large common proper motion makes it almost certain that the stars are physically related. If the brighter star populates the Hyades main sequence the parallax is 0″039, the tangential velocity is 75 km/sec and the fainter component is a RSL star with $M(I) = +11.6$ at $(R-I) = +0^m46$, making it nearly identical to LTT 2236 in Table I.

References

Altena, W. F. van: 1969, *Astron. J.* **74**, 2.
Eggen, O. J.: 1969a, *Astrophys. J. Suppl.* **19**, 31.
Eggen, O. J.: 1969b, *Astrophys. J.* **158**, 1109.
Eggen, O. J.: 1970a, *Astrophys. J.* **159**, 945.
Eggen, O. J.: 1970b, *Astrophys. J.*, in press.
Eggen, O. J. and Greenstein, J. L. :1965, *Astrophys. J.* **141**, 83.
Greenstein, J. L. and Eggen, O. J.: 1966, in *Vistas in Astronomy* (ed. by A. Beer), Pergamon Press, New York.
Joy, A. H.: 1947, *Astrophys. J.* **105**, 96.
Luyten, W. J.: 1957, *Catalogue of 9867 stars in the Southern Hemisphere with Proper Motion Exceeding 0″.2 Annually*, Lund Press, Minneapolis.

3. PARALLAXES OF WHITE DWARFS

K. Aa. STRAND

U.S. Naval Observatory
Washington, D.C., U.S.A.

The U.S. Naval Observatory program on trigonometric stellar parallaxes with the 61-inch Astrometric Reflector has been in progress since 1964.

The program has been devoted entirely to stars fainter than visual magnitude $12^{m}.5$ for which there were reasonable assurances, from their proper motions, of their being intrinsically faint and, hence, either dwarfs, faint main sequence stars or subdwarfs. The parallaxes for the first 100 stars will appear in *Publ. U.S. Naval Observatory* **XX**, Part III. Part IIIA, by R. K. Riddle, will describe the astrometric data; Part IIIB will contain the photoelectric colors and magnitudes in the U, B, V system, as determined by J. B. Priser; and Part IIIC will have a general discussion of the data in relation to the H-R Diagram, by K. Aa. Strand and R. K. Riddle.

In this group of 100 stellar parallaxes, 18 stars are white dwarfs.

Data for a second group of 50 stars are nearing completion and include an additional 12 white dwarfs.

On the average, each parallax is based on 45 plates taken over a period of more than 3 years leading to a parallax determined with an internal mean error between $\pm 0''.003$ and $\pm 0''.004$. In view of the high stability of the optical system, the internal consistency of the material, the large number of plates involved, and the high accuracy of the auto-centering feature of the measuring machine, an external mean error of $\pm 0''.010$ is a conservative estimate and could possibly be as low as $\pm 0''.007$.

The H-R Diagram, as derived from the above, is shown in Figure 1. The large filled circles indicate stars with parallaxes greater than or equal to $0''.045$ ($\Delta M_V = \pm 0^{m}.6$ or less). Open circles indicate parallaxes less than $0''.045$ but larger than or equal to $0''.025$ ($\Delta M_V \sim \pm 1^{m}.0$), and the dots indicate parallaxes with values less than $0''.025$ ($\Delta M_V > 1^{m}.0$). The values in the parenthesis indicate the corresponding error in absolute magnitude based upon an external mean error of $\pm 0''.010$ of the parallax. The Johnson-Morgan (1953) standard main sequence is indicated by the drawn curve (Vyssotsky, 1963).

As seen from the diagram, the reddest white dwarfs for which parallaxes were obtained have a $B-V$ of 1.0, leaving a substantial gap between the main sequence and the white dwarf branch. The star G7-17, which was considered a possible white dwarf by Eggen and Greenstein (1965) near the intersection of the normal M-type dwarf sequence, was included in the program. It turned out to have a very small parallax ($0''.015 \pm 0''.005$ (m.e.)) and not a parallax of $0''.068$ ($M_V = 14^{m}.7$), as assumed by the above authors. If the parallax determined at the Naval Observatory is correct, the star lies above the red end of the main sequence.

The bluest star in the program is G191-B2B, with a parallax of $0''.021 \pm 0''.003$ (m.e.),

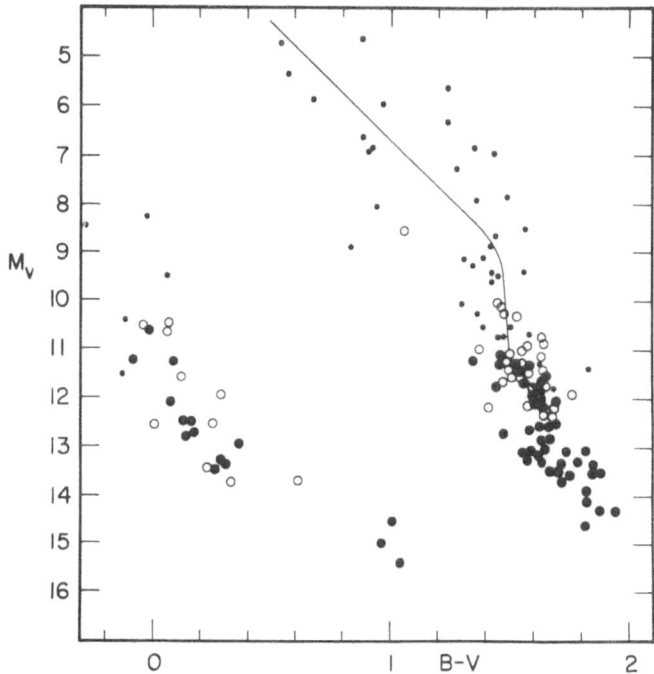

Fig. 1.   The H-R diagram for stars for which parallaxes and photometric data were obtained at
the U.S. Naval Observatory. The key to the symbols is described in the text.

leading to $M_V = 8^m_.4$. The spectrum of the star, according to Eggen and Greenstein
(1967), is sdO or DAwk. Its annual proper motion is $0^{''}_.093$, leading to a peculiar
tangential velocity of only 2 km/sec. The star is the fainter companion to G191-B2A,
which has been classified as a sdK4p by the same authors. The parallax determined
here places the star on the main sequence.

The other bright white dwarf is AC82° 3818 with a parallax of $0^{''}_.011$, which is a
preliminary value based on 19 plates only. If the parallax is confirmed by further
observations, the absolute magnitude of this star will remain uncertain from trigonometric data. Eggen and Greenstein (1965) have classified it as DA, and estimate the
absolute magnitude to be either $11^m_.0$ or $12^m_.5$ with corresponding parallaxes of $0^{''}_.040$
and $0^{''}_.079$.

In Figure 2 we find plotted in the M_V, U−V diagram the white dwarfs shown in
Figure 1, using the same symbols.

In addition are plotted stars (+) for which trigonometric parallaxes greater than
$0^{''}_.040$ exist from other sources. Also plotted are the white dwarfs in the Hyades group
(□) and in the Praesepe Cluster (◇), listed in the paper by Eggen and Greenstein
(1967).

This represents the inventory of absolute magnitudes of white dwarfs known at this
time, as determined from astrometric data.

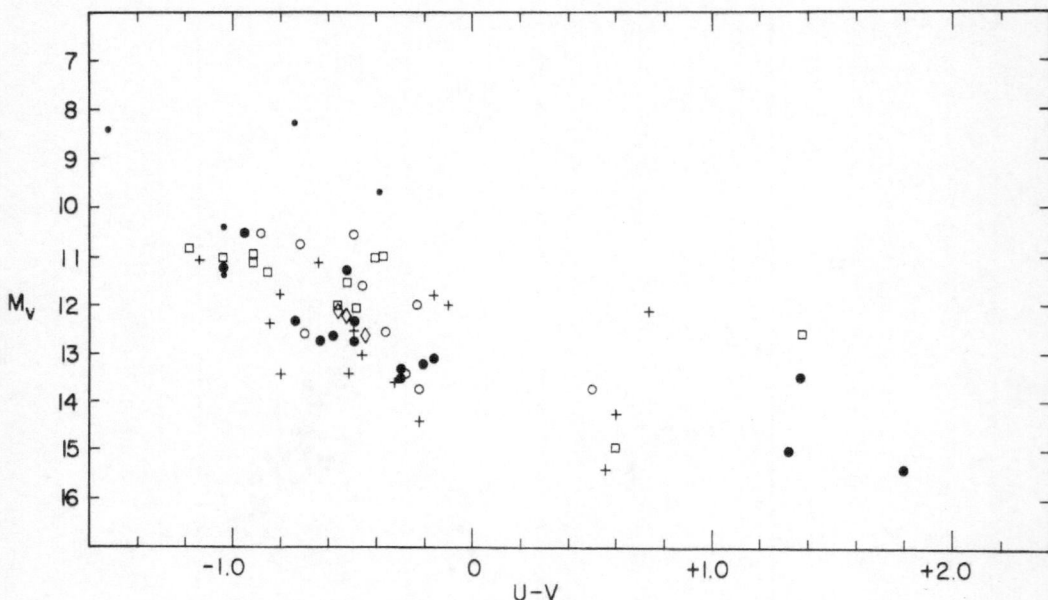

Fig. 2.   A M_V, U − V diagram of white dwarfs. The key to the symbols is described in the text.

Acknowledgement

It should be noted that the Naval Observatory team has substantially increased this inventory and, hopefully, will continue to do so in the years to come.

References

Eggen, O. J. and Greenstein, J. L.: 1965, *Astrophys. J.* **141**, 83.
Eggen, O. J. and Greenstein, J. L.: 1967, *Astrophys. J.* **150**, 927.
Johnson, H. L. and Morgan, W. W.: 1953, *Astrophys. J.* **117**, 313.
Vyssotsky, A. N.: 1963, *Basic Astronomical Data* (ed. by K. Aa. Strand), University of Chicago Press, Chicago, p. 192.

4. TRIGONOMETRICAL PARALLAXES OF LB 3303
AND LB 3459

J. CHURMS

Royal Observatory, Cape, South Africa

and

A. D. THACKERAY

Radcliffe Observatory, Pretoria, South Africa

1. LB 3303

Radcliffe spectra [1] of this star in 1961 confirmed Luyten's suspicion on the basis of colour and proper motion that this object is a bright white dwarf. The profile of $H\gamma$ was found to be closely similar to that of o_2 Eri B. Hill and Hill [2] have published photometry yielding $V = 11.40$. The only known white dwarfs brighter than this are Sirius B, o_2 Eri B, Procyon B, Feige 34 and CD $-38°$ 10980.

A trigonometrical parallax has been determined by one of us (J.C.) at the Royal Observatory, Cape. The result from 28 plates using 5 comparison stars is

Relative $\quad \pi_t = 0\rlap{.}''067 \pm 0.011$ (p.e.)

or following Binnendijk [3]

Absolute $\quad \pi_t = 0\rlap{.}''070 \pm 0.011$

The resulting absolute magnitude is

$$M_v = +10.6 \pm 0.3$$

This result is derived without the systematic correction of $-0\rlap{.}''002$ to Cape parallaxes recommended by Jenkins [4] since Strand [5] has advised against such Yale-Jenkins system of corrections. (If the Yale-Jenkins precepts were followed the derived absolute parallax would become $+0.065 \pm 0.014$.)

The derived M_v ($+10.6$) makes the star rather luminous for a white dwarf, but the observations are consistent with $M_v = +10.8$, the value for W 485 (with very similar colours) and the hotter object W 1346 (Eggen and Greenstein [6]). The star appears to be 0.8 ± 0.3 brighter than the Eggen-Greenstein [6] mean relation

$$M_v = 11.65 + 0.85 (U - V)$$

Nevertheless, the parallax provides the final confirmation without doubt that the star is a relatively near white dwarf.

2. LB 3459

This is an extremely blue galactic star superposed on the Large Magellanic Cloud, discovered by Luyten on the basis of its proper motion. Radcliffe spectroscopy [7] confirmed its membership of the galactic foreground and also the variation in its

Luyten (ed.), White Dwarfs, 21–23. All Rights Reserved.

spectrum already noted by Miss Cannon although only the first Radcliffe spectrum showed strong HeI lines, MgII 4481 and a peculiarly strong Ca$^+$ K line which cannot be interstellar. Spectroscopic monitoring has been continued and of 12 Radcliffe spectra taken between 1958 and 1968 all but the first still show weak helium lines.

The light also appears to be constant.

Although the spectrum does not suggest that of a white dwarf, the significant proper motion meant that a measurement of parallax was desirable and it was put on the Cape programme. The result from 27 plates using 5 comparison stars is given below, together with the Binnendijk correction from relative to absolute.

Relative $\pi_t = -0''013 \pm 0.008$ (p.e.)

Absolute $\pi_t = -0''010 \pm 0.008$

If we assume that the error of this determination is less than 4 times its probable error, we derive a minimum distance of 45 parsec, and with the Hills [2] photometry, $M_v \leq +7.8$. The star cannot be regarded as a white dwarf on the basis of its spectrum or absolute magnitude. On the other hand the fact that the star exhibits appreciable proper motion as discovered by Luyten sets an upper limit to the luminosity. A newly determined Cape proper motion of $0''038$, yields a tangential motion T for various assumed distances as set out in Table I.

TABLE I

r (parsec)	M_v	T (km/s)	z (parsec)
45	7.8	8.3	24
180	4.8	33	96
360	3.3	66	192
720	1.8	132	384

With a small radial velocity (certainly less than 50 km/sec) we can say that

$$+1.8 < M_v \leqslant +7.8$$

Thus LB 3459 appears to be an unusual object lying between the hot subdwarfs and the white dwarfs. HeII 4686 appears weakly in absorption on some Radcliffe spectra and thus a revised classification 'OBp' is perhaps appropriate.

Table II summarises information about the two stars. Successive columns give (1) star designation from various catalogues, (2) galactic coordinates, (3) V, B−V, U−B from the Hills' photometry, (4) Spectral classification and derived M_v, (5) Trigonometrical parallax from this determination, (6) newly determined Cape proper motion (by J.C.), (7) previously published (CPC 50) proper motion.

Acknowledgements

We are indebted to Mrs. B. Brown, Mrs. M. C. Coetzee, and Miss A. J. Goedhals who performed the measures of the trigonometrical parallaxes at the Royal Observatory, Cape.

TABLE II

Star	l^{II} b^{II}	V B − V U − B	Sp	π_t M_v	(p.e.)	Relative $\mu\alpha$ $\mu\delta$	(p.e.)	CPC50 $\mu\alpha$ $\mu\delta$
LB 3303 − 69°177 EG 21	286.2 − 43.7	11.40 + 0.05 − 0.55	DA	+ 0″.070 + 10.6	(0.011) (0.3)	+ 0″.038 − 0.104	(0.004) (0.004)	
LB 3459 − 69°389 269696 (HDE)	280.5 − 32.2	11.13 − 0.27 − 1.10	OBp p	− 0″.010 ⩽ + 7.8	(0.008)	− 0″.008 + 0.038	(0.004) (0.004)	− 0.002 (0.004) + 0.044 (0.004)

References

[1] Thackeray, A. D.: 1961, *Monthly Notes Astron. Soc. S. Afr.* **20**, 40.
[2] Hill, P. W. and Hill, S. R.: 1966, *Monthly Notices Roy. Astron. Soc.* **133**, 205.
[3] Binnendijk, L.: 1943, *Bull. Astron. Inst. Neth.* **10**, 9.
[4] Jenkins, L. F.: 1952, *General Catalogue of Trigonometric Stellar Parallaxes.*
[5] Strand, K. Aa.: 1963, 'Basic Astronomical Data', *Stars and Stellar Systems* **III**, 55.
[6] Eggen, O. J. and Greenstein, J. L.: 1965, *Astrophys. J.* **141**, 83.
[7] Feast, M. W., Thackeray, A. D., and Wesselink, A. J.: 1960, *Monthly Notices Roy. Astron. Soc.* **121**, 383.
[8] Luyten, W. J.: 1959, *A Search for Faint Blue Stars* **XVII**, Minneapolis.

5. THE IDENTIFICATION OF WHITE DWARF SUSPECTS IN THE LOWELL PROPER MOTION PROGRAM

H. L. GICLAS

Lowell Observatory, Flagstaff, Ariz., U.S.A.

During the course of the Lowell proper motion survey that has been in progress since 1957, an effort has been made to identify the very blue stars and more recently the extremely red ones. The proper motions are detected and measured directly by projection on the calibrated grid of the blink microscope. The regular program catalogs all motions found $> 0\overset{''}{.}26$/year and fainter than the eighth magnitude. From this portion of the program, we have identified 179 white dwarf suspects, 48 of color class -1 and 131 of color class 0, which have been published in the *Lowell Bulletins* under the title of 'Lowell Proper Motions II through XV'. These lists cover the entire northern hemisphere, and a few regions down as far as $-10°$ South, and contain 10382 different individual stars. Finding charts are provided for each object.

In order to realize a high degree of completeness at the lower adopted limit of motion, an attempt is made to mark all stars showing detectable motion; this includes many motions as small as $0\overset{''}{.}10$/yr. If, when comparing a blue with a red plate for color estimates, any of these marked stars or any others in the vicinity that show no appreciable motion appear to be very blue, they are noted. It is from these candidates that the 'GD' lists of 'White Dwarf Suspects' are prepared. Three such lists have been published (*Lowell Bulletin*, Nos. 125, 141, and 153) and contain 562 white dwarf suspects of various degrees of certainty. The most recent list (*Lowell Bulletin*, No. 153) of 155 GD stars has just been completed as a part of this report, and completes the list of white dwarf suspects of small motions from $+50°$ to the north pole.

In order to assess the true number of new white dwarfs contained in these lists, we refer to the papers of Eggen and Greenstein [3, 4, 5, and 6] where, either from the photometry by Eggen or from spectra taken by Greenstein, they have found that essentially all stars of color class -1 and 60% of color class 0 are white dwarfs (7). Therefore, one may expect to verify 145 white dwarfs from the regular program, and 335 from the GD lists for a total of 480 new ones. This does not include several hundred additional ones that might have Luyten, Tonanzintla, or other previous references that are included in both lists.

The number of stars in each color and motion class for the GD stars is shown in Figure 1. Most are in color class 0, and about equally distributed between our EPM (Estimated Proper Motion) 1 and 2, the small motions.

Plotting measured small proper motions from several sources, mostly Luyten, against our EPM, we find the relation shown in Figure 2. These are compared in Table I with the values used by Eggen and Greenstein [5], denoted EG in the table, to check the luminosity classes from the tangential motion. They differ slightly, but are based on perhaps twice as many small motions as were available to them.

Luyten (ed.), White Dwarfs, 24–31. *All Rights Reserved.*

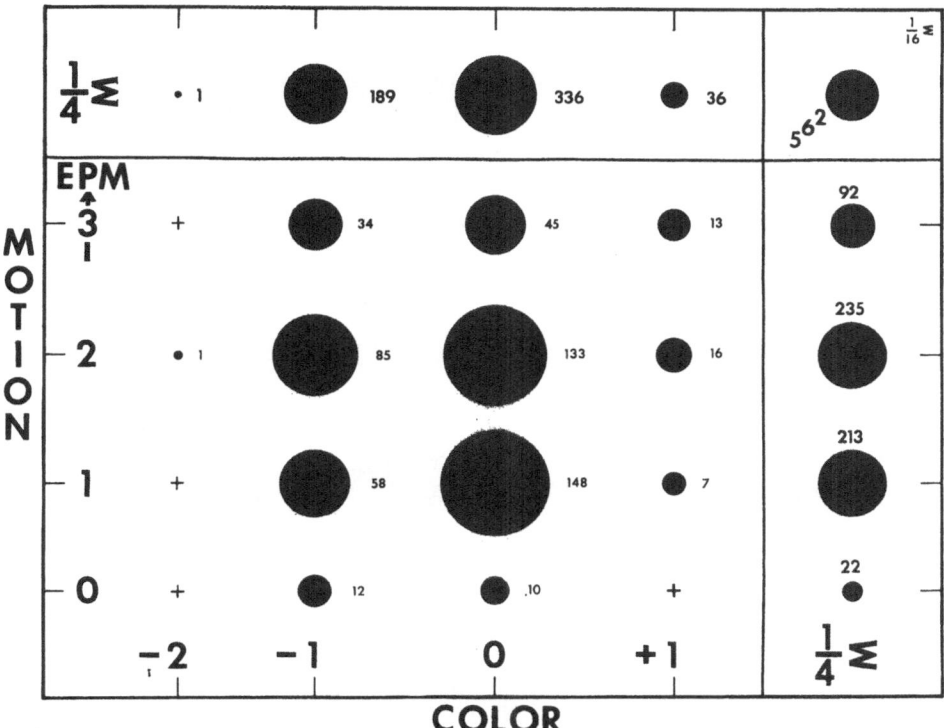

Fig. 1.   The number of GD Stars in each color and EPM (Estimated Proper Motion) class.

To place a quantitative value on our color classes for the GD suspects is a little more involved if we want to define a reasonable narrow range. If we plot the UBV color of confirmed white dwarfs against our color classes 0 and −1, we find the distribution shown in Figure 3. The mean value for the color class taken from this figure is shown in Table II. Examination of the U−B range of colors for the 0 class stars in Figure 3 shows a marked secondary distribution around U−B = −0.14 mag. This group of objects was identified as stars with the smallest proper motion.

TABLE I

Lowell class		Average values	
		Motion	Deviation
EPM 3	L =	0″.213	±0″.037
	EG =	0″.246	±0″.049
EPM 2	L =	0″.168	±0″.036
	EG =	0″.167	±0″.034
EPM 1	L =	0″.107	±0″.026
	EG =	0″.112	±0″.025

As evidenced by the greater range of color for 0 class dwarf suspects, one would
expect to find a more diverse assortment of objects to be included in that color class.
To see if there may be a sorting of different types of degenerate stars dependent upon
apparent tangential velocity, the B−V and U−B has been plotted for each EPM.
This is shown in Figures 4 and 5, and the color equivalents derived from them shown

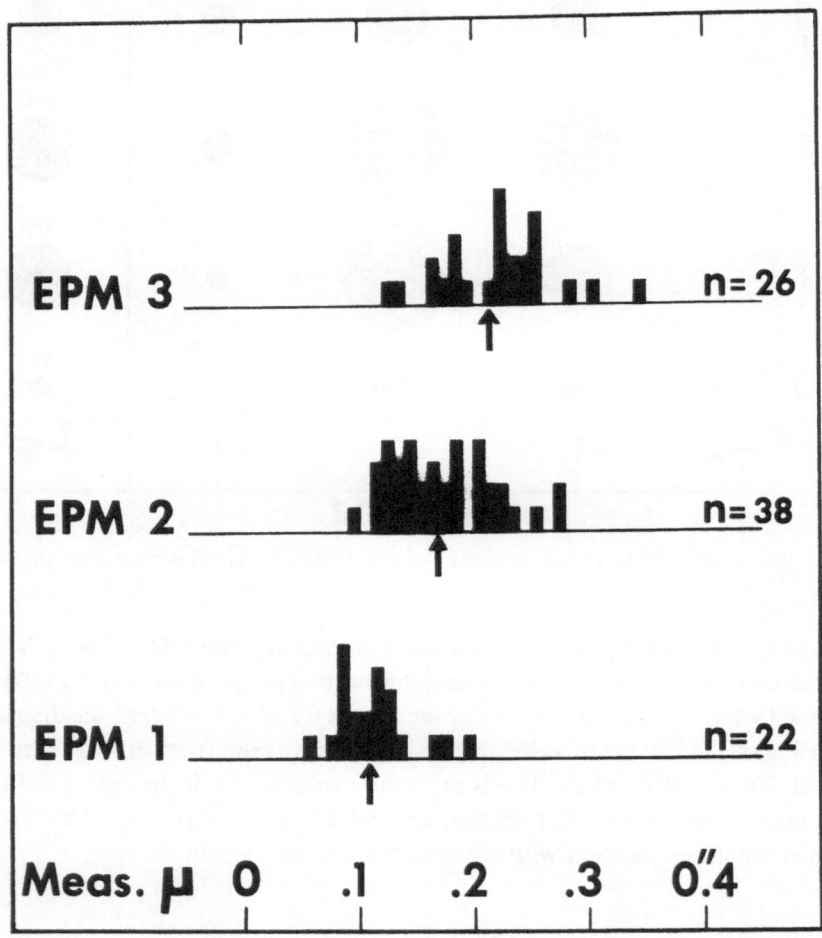

Fig. 2.   Estimated Proper Motion, EPM, vs measured motion of Luyten and others.

in the last three columns of Table II. A dual value for the U−B of 0 color and EPM 1
class is given; one based on all stars, U−B = −0.60, and one, U−B = −0.14, indicated
by an asterisk in Table I, for the subdwarfs. It is apparent from this that a preponder-
ance of subdwarfs have been identified among the 0 color class with the smallest
proper motion.

Greenstein [6], out of 86 GD stars classified spectroscopically, found 54 were early type white dwarfs, and the greatest difference between them and the program stars with large proper motion is the higher frequency of DB, or helium rich stars. Thirty-two of this sample were not white dwarfs; 22 of the 32 were horizontal branch stars or yellow subdwarfs that appear blue on our plates because of their large ultra-

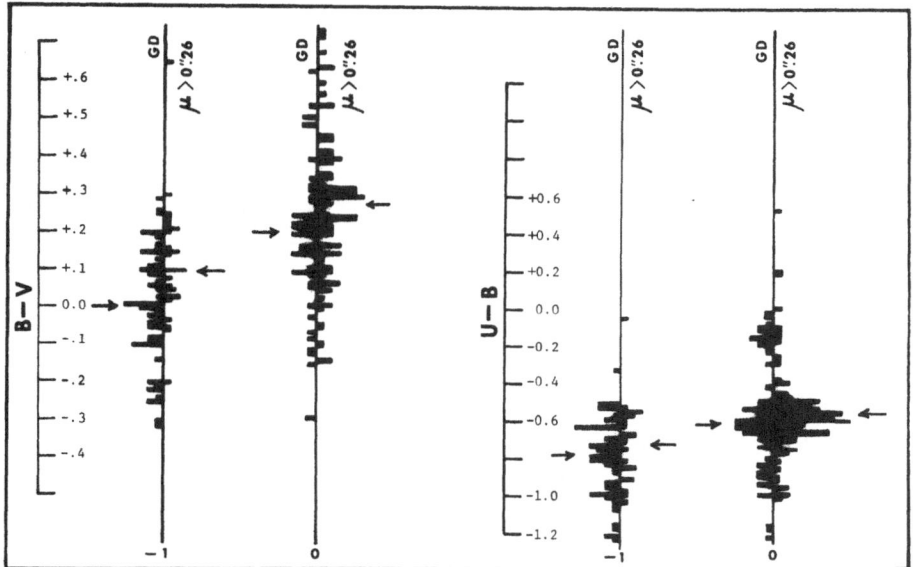

Fig. 3. B − V and U − B vs Lowell color class 0 and − 1 for the confirmed GD and $\mu > 0''.26$/yr stars.

TABLE II

Figure 3				Figure 4 and 5		
	C	GD	$\mu > 0''.26$	EPM		
				1	2	3
B − V	− 1	0.00	+ 0.10	0.00	0.00	+ 0.13
	0	+ 0.20	+ 0.28	+ 0.35	+ 0.22	+ 0.18
	− 1	− 0.78	− 0.72	− 0.78	− 0.78	− 0.65
U − B	0	− 0.60	− 0.55	− 0.60	− 0.55	− 0.65
				− 0.14*		

violet excess. A detailed discussion of the GD stars can be found in Greenstein's paper, 'The Lowell Suspect White Dwarfs' [6].

While the value for the position angle of the GD's is admittedly very rough, being estimated to the nearest 5°, an attempt to detect any excess number of these small motion objects moving toward selected convergents was investigated. The number of these stars moving toward each of the following three convergents were identified by the residual angle [9] technique previously described:

Hyades 6^h08^m + 9°00′
61 Cygni   6 37   + 0 30
Ursa Major   20 21   − 54 00

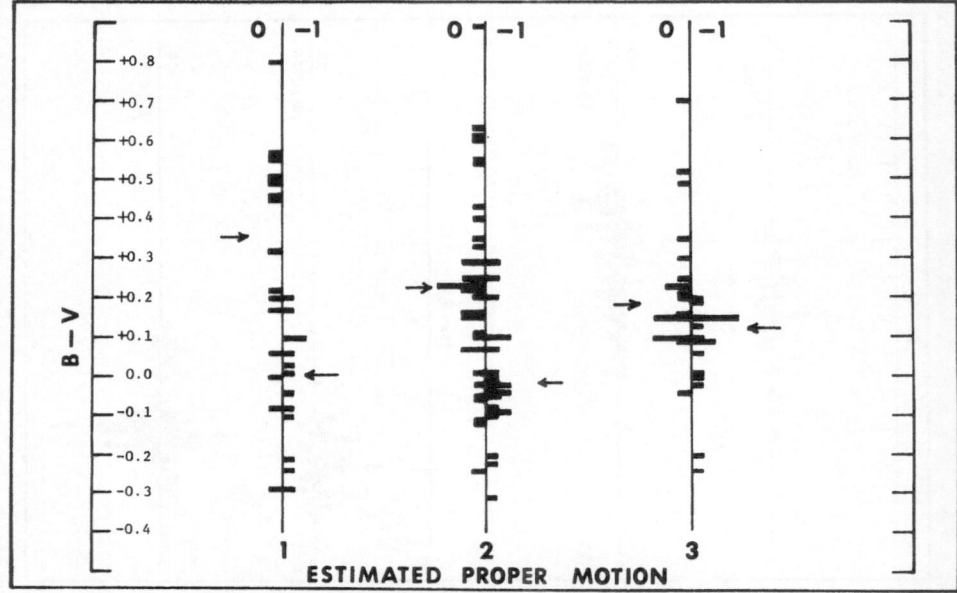

Fig. 4.   B − V vs EPM for the Lowell GD dwarfs of color class 0 and − 1.

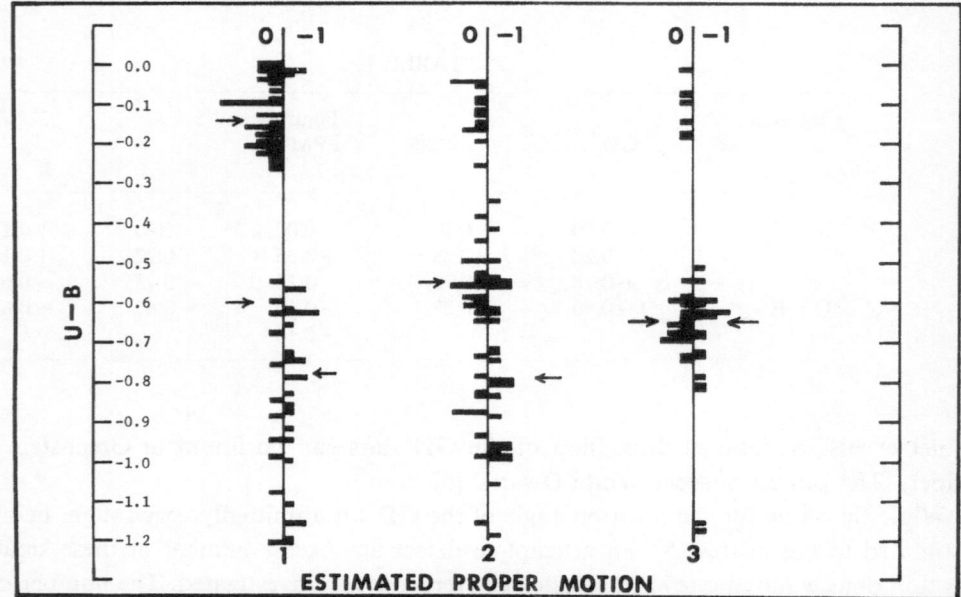

Fig. 5.   U − B vs EPM for the GD dwarfs of color class 0 and − 1.

This procedure maximizes the separation of motion toward a convergent from the nearby Kapteyn Drift.

It was only in the case of the Hyades Group that there appeared to be an excess number of stars moving toward a convergent. The plot of the distribution of the residual angles is shown in Figure 6, where the Kapteyn Drift I has been taken to be $7^h09^m - 13°00'$, the position of the weighted direction of streaming determined from our own data [9]. The position of the Hyades convergent is at 0°, abscissa, of the figure, and the number of stars moving in that direction is the ordinate. The normal number of background stars within 40° of the direction of the Hyades convergent, determined by symmetry of the curve around Drift I, should be 21 stars. Actually, the total number of stars found under the secondary, dotted, hump is 32, therefore there appears to be an excess number of 11 stars travelling towards this convergent point. These 32 stars are listed in Table III. UBV photometry has been done by Eggen [10] for 15 of these, and are identified by the E10 appearing in the column headed References. Accurate proper motion and radial velocities will be needed to verify which ones of this list might be true Hyades Group members.

The Lowell proper motion program has been completed in the northern hemisphere, and the data is being prepared to print out a catalog ordered in right ascension, after averaging duplicate measures from overlapping plates. Columns will be provided for UBV photometry and reference to parallaxes where they exist. It is planned to continue the program to at least $-30°$ South declination as first epoch plates taken

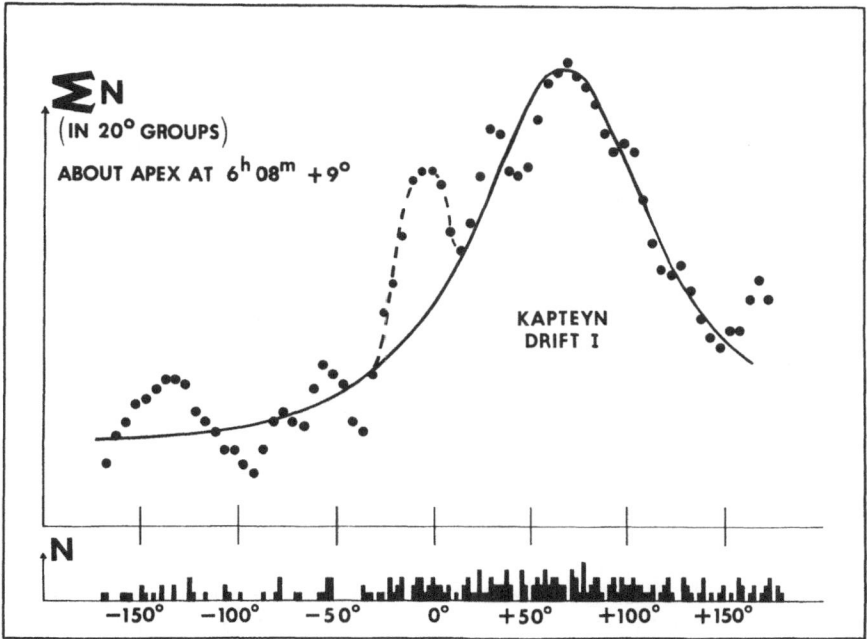

Fig. 6. Distribution of residual angle. Abscissa, position angle from Hyades group Apex at 0°, ordinate, number of stars.

with the 13-inch telescope are available to declination −40°. We are now in the process of moving this telescope to a dark sky site 12 miles SE of its present city location so that second epoch plates can be made to match the sky background of the first.

Some supplementary things we would like to do are to re-examine the first 60-plate

TABLE III

List containing eleven possible Hyades group members

Number	R.A.	Dec.	EPM	P.A.	Mag.	Color	References
GD 6	0ʰ28ᵐ07ˢ	+ 16°44′	1	80°	16.0	+ 1	
9	58 29	− 4 27	1	80	15.0	− 1	PHL 940, E10
19	1 50 26	+ 9 34	2	80	14.0	0	E10
31	2 31 37	− 5 24	3	70	14.0	0	PHL 1358, E10
51	3 47 54	− 13 44	2	60	15.0	0	E10
52	3 48 48	+ 33 58	2	120	15.0	− 1	E10
436	5 17 39	+ 60 25	1	165	15.5	0	
437	18 59	+ 71 37	2	170	13.5	0	LP33-137
440	44 34	+ 76 15	2	175	14.5	0	
75	6 28 38	+ 47 39	3	190	14.0	0	E10
77	6 37 26	+ 47 47	3	190	15.0	− 1	E10
292	41 26	+ 53 35	1	195	15.0	0	
84	7 14 23	+ 45 53	3	205	15.5	− 1	E10
111	10 02 46	+ 43 02	1	260	16.0	− 1	E10
463	16 13	+ 63 08	1	250	13.5	0	
464	23 26	+ 68 02	1	255	16.5	+ 1	
125	52 01	+ 27 22	2	270	14.0	− 1	Ton 556, E10
466	11 02 25	+ 74 52	2	255	15.0	0	
127	02 42	+ 0 32	3	280	16.0	0	E10
135	17 02	− 2 22	3	280	15.0	0	E10
308	22 59	+ 42 40	2	270	16.0	− 1	
146	12 00 34	+ 21 18	2	280	16.0	0	E10
164	14 13 38	+ 28 21	1	300	16.0	0	E10
343	15 19 00	+ 63 40	2	320	16.0	− 1	LP68-60
187	38 36	+ 33 18	3	320	16.0	0	LTT 14655
194	50 12	+ 18 19	2	310	14.5	0	LTT 14705, E10
359	17 06 05	+ 35 36	2	340	16.0	− 1	
364	45 13	+ 39 53	1	360	16.5	− 1	
396	21 43 16	+ 35 18	3	60	15.5	− 1	
240	22 40 09	− 4 30	3	80	16.0	− 1	PHL 380, Feige 106
562	23 50 41	+ 74 37	1	85	15.0	0	
255	58 36	+ 20 38	2	80	15.5	+ 1	

regions for white dwarf suspects, and for extremely red stars. In continuing the program into the southern hemisphere it will be possible to match first epoch plates around the south galactic pole made in 1934, and to do a proper motion study in that area in depth. Perhaps the most useful addition to our program would be to do the UBV photometry of the dwarf suspects and other interesting objects; however, additional support would have to be found for this.

Acknowledgments

It is a pleasure to acknowledge the contributions of my colleagues, Mr. Robert Burnham, Jr. and Mr. Norman G. Thomas, who have been with the Lowell proper motion program for the past 12 and 11 years respectively. This program has been supported by National Science Foundation grants which are gratefully acknowledged.

References

[1] Giclas, H. L., Burnham, Jr., R., and Thomas, N. G.: 1957–1970, 'Lowell Proper Motions' I through XV; *Lowell Observatory Bulletin*, Nos. 89, 102, 112, 120, 122, 124, 129, 132, 136, 138, 140, 144, 150, 151 and 152.
[2] Giclas, H. L., Burnham, Jr., R., and Thomas, N. G.: 'A List of White Dwarf Suspects', I, II and III; *Lowell Observatory Bulletin*, Nos. 125, 1965; 141, 1967 and 151, 1970.
[3] Eggen, O. J. and Greenstein, J. L.: 1965a, *Astrophys. J.* **141**, 83.
[4] Eggen, O. J. and Greenstein, J. L.: 1965b, *Astrophys. J.* **142**, 925.
[5] Eggen, O. J. and Greenstein, J. L.: 1967, *Astrophys. J.* **150**, 927.
[6] Greenstein, J. L.: 1969, *Astrophys. J.* **158**, 281.
[7] *Annual Report of the Director 1964–1965*, Mt. Wilson and Palomar Observatories.
[8] *Annual Report of the Director 1965–1966*, Mt. Wilson and Palomar Observatories.
[9] Giclas, H. L.: 1969, *Low Luminosity Stars*, Kumar, p. 28.
[10] Eggen, O. J.: 1968, *Astrophys. J., Suppl. Series* **16**, 97, No. 143.

6. AN ASTROMETRIC STUDY OF VAN MAANEN'S STAR

P. VAN DE KAMP

Swarthmore College, Swarthmore, Pa., U.S.A.

Van Maanen's star (1950: 0^h46^m5, $+5°09'$) is a white dwarf (visual magnitude 12.4, spectrum DF) with a proper motion of $2''95$ in $155°5$ and parallax $0''234$. An early determination of radial velocity yielded $+238$ km/sec (Adams and Joy, 1926). It was suggested by Russell and Atkinson (1931) that van Maanen's star might show a very large Einstein shift of at least $+700$ km/sec. Oort (1932) pointed out that the observed value of $+238$ km/sec is incompatible with the theory of galactic rotation: this would support the Einstein shift interpretation.

Meanwhile radial velocity and Einstein shift have been drastically revised. According to Greenstein (1954) radial velocity results appear to depend on dispersion and exposure time; low dispersion spectra yield a range of $+21$ to $+216$ km/sec. Observations by Greenstein and Trimble (private communication) yield $+54$ km/sec; neutral iron lines yield about -50 km/sec, while a mass of $0.68 \odot$ and radius $\frac{1}{78} R_\odot$ would yield an Einstein shift of $+34$ km/sec.

Oort (1932) proposed that a determination of secular change in proper motion would eventually provide a direct geometric determination of the radial velocity. The yearly secular perspective acceleration is given by

$$\Delta\mu = - 2''05 \times 10^{-6}V\mu p \qquad (1)$$

Conversely the radial velocity is given by

$$V = - 4.88 \times 10^5 (\Delta\mu/\mu p) \text{ km/sec} \qquad (2)$$

or, for van Maanen's star

$$V = - 2.08 \times 10^6 (\Delta\mu/\mu) \text{ km/sec} \qquad (3)$$

A first attempt at determining the acceleration is presented here. van Maanen's star was put on the observing program of the Sproul 24-inch refractor ($1 \text{ mm} = 18''87$) in 1937. Results for parallax and proper motion from plates up to and including 1948 have been published (van de Kamp and Lippincott, 1949). The additional material obtained since 1948, together with a number of earlier plates, was measured by Mrs. Betty Kuhlman on the St. Clair-Kasten machine, and again reduced to the standard frame based on three reference stars. The distribution of the material is uneven, which is partly explained by the change in exposure times which averaged 50 min in 1937 and 1938, about 20 min from 1939 to 1944, and gradually decreased to the current value of about 4 min.

A solution for parallax, proper motion and acceleration was made from 281 plates with 511 exposures and a total night weight 174 taken on 137 nights spread over the interval 1937.0–1970.0. The nightly mean positions are represented by

$$X = c_x + \mu_x t + q_x t^2 + \pi P_\alpha$$
$$Y = c_y + \mu_y t + q_y t^2 + \pi P_\delta$$

where t is counted in years from 1950.000. A least squares solution combining x and y gives

$$\mu_x = + 1.''2214 \qquad \pm 0.''0009 \quad \text{(p.e.)}$$
$$\mu_y = - 2.6843 \qquad 0.0009$$
$$\pi = + 0.2231 \qquad 0.0042$$
$$q_x = + 0.000052 \qquad 0.000032$$
$$q_y = - 0.000162 \qquad 0.000032$$
$$\text{p.e. } 1 = \pm 0.^{mm}00187 = \pm 0.''0352$$

Separate solutions yield

$$\pi_x = + 0.''2294 \qquad \pm 0.''0046$$
$$\pi_y = + 0.1879 \qquad 0.0104$$
$$\text{p.e. } 1_x = \pm 0.^{mm}00188 = \pm 0.''0355$$
$$\text{p.e. } 1_y = \pm 0.00182 \quad = \pm 0.0343$$

The residuals from the least squares solution indicate a possible perturbation with a period of some two decades and an amplitude too small to have a sensible effect on the measured acceleration.

The observed acceleration is affected by a spurious acceleration caused by the proper motions of the reference stars and their changing influence, with time, on the reduced position of the central star (van de Kamp, 1935). We were fortunate in being able to measure the proper motions of the three reference stars on a background of 60 stars, using four pairs of plates with an average interval of 28.6 years. The probable errors of the resulting proper motions are $\pm .''0015$ and $\pm .''0012$ in x and y, respectively.

The relative proper motions and annual dependence changes for the reference stars are as given in Table I.

TABLE I

	μ_x	μ_y	ΔD
1	$+ 0''.0210$	$- 0''.0295$	$- 0.00348$
2	$-\ \ \ 160$	$-\ \ \ 137$	$+\ \ \ 260$
3	$-\ \ \ 23$	$+\ \ \ 44$	$+\ \ \ 88$

The corresponding corrections for spurious acceleration are:

$$+ 2[\Delta D \mu_x] = - 0.''000234 \pm 0.''000007$$
$$+ 2[\Delta D \mu_y] = + 0.000132 \pm 0.000005$$

We thus obtain the values as given in Table II for the acceleration components and the radial velocity using formula (3).

TABLE II

	x		y	
Observed acceleration	$+0''.000104$	$\pm 0''.000064$	$-0''.000326$	$\pm 0''.000064$
Correction for spurious acceleration	$-0''.000234$	0.000007	$+0''.000132$	0.000005
True perspective acceleration	-0.000130	0.000064	-0.000194	0.000064
Radial velocity	$+220$	± 110 km/sec	$+150$	± 50 km/sec

Assigning relative weights of 1 and 5 to the determinations in x and y, we find a resultant value of $+160\pm45$ km/sec for the radial velocity determined by the present geometric method. Although this value must be considered provisional, it appears to exclude very large values either a large positive value for the radial velocity or the Einstein red shift.

The present result is not anywhere near as accurate as the corresponding determination recently made for Barnard's star from some 3500 plates taken over the interval 1916–1919 and 1938–1969, resulting in a probable error of only 2.6 km/sec (van de Kamp, 1970). In that study the obtainable accuracy was limited by the accuracy of the spurious acceleration, which had the much larger probable error of $0''.00003$, due to the large annual dependence changes. For van Maanen's star the spurious acceleration is very accurately determined. In the not too distant future we shall be able to make a much more accurate geometric determination of the radial velocity of van Maanen's star. Observational material, distributed uniformly in time, yields an acceleration whose weight increases with the fifth power of the time interval. In the present case the material has a far from uniform distribution in time, observations between 1953 and 1962 being very scarce. With the current short exposure time extensive future material appears to be assured and a vast improvement of the perspective acceleration, and hence the radial velocity, is expected within the decade.

References

Adams, W. S. and Joy, A. H.: 1926, *Publ. Astron. Soc. Pacific* **38**, 122.
Greenstein, J. L.: 1954, *Astron. J.* **59**, 322.
Greenstein, J. L. and Trimble, V. L.: 1967, *Astrophys. J.* **149**, 293.
Oort, J. H.: 1932, *Bull. Astron. Inst. Neth.* **238**, 287.
Russell, H. N. and Atkinson, R. E.: 1931, *Nature* **137**, 661.
van de Kamp, P.: 1935, *Astron. J.* **44**, 73.
van de Kamp, P. and Lippincott, S. L.: 1949, *Astron. J.* **55**, 16.
van de Kamp, P.: 1970, IAU Colloquium No. 7 on Proper Motions.

7. THE WHITE DWARFS IN THE CATALOGUE OF
NEARBY STARS OF 1969*

W. GLIESE

Astronomisches Rechen-Institut, Heidelberg, Germany

The Catalogue of Nearby Stars (Gliese, 1969) contains 1529 single stars and systems with altogether 1890 components. It includes 54 white dwarfs with apparent magnitudes from 8.5 to 17; namely 38 single stars and 16 degenerate stars which are components of binaries or triples. Only 48 of these objects have parallaxes larger than 0".050. But our knowledge is very incomplete.

The distances of 33 of these white dwarfs have been determined trigonometrically with probable errors (p.e.) between 2% and 30%, 4 distances are from the photometric parallaxes of the companions, one star is a Hyades member, another belongs to the Wolf 219 group (group parallax), and for 15 objects the absolute magnitudes have been taken from the $(M_V, U-V)$ relations by Eggen and Greenstein (1965).

The absolute magnitudes of 18 degenerate stars are known with probable errors smaller than $\pm 0^{m}.21$, but only for 8 of these objects have U, B, V photometry and space velocities also been determined. This demonstrates the incompleteness of our knowledge.

Nevertheless I shall try to point to some questions which may be indicated by this material.

In the considerations concerning velocities we exclude the DAss star LP 9-231 whose space velocity is nearly 350 km/sec.

(a) *Radial Velocities.* The radial velocities of 18 of these degenerate stars have been measured. After elimination of the standard solar motion the mean RV is +45 ±4.5 km/sec (p.e.), in good agreement with the mean Einstein effect of 51 km/sec derived by Greenstein and Trimble (1967). The space velocities in the Nearby Star Catalogue have been computed after correcting the measured RV by −51 km/sec.

The comparison of the measured RV's of o^2 Eri B and of CoD −37°10500 B with that of their companions gives differences of +22 resp. +56 km/sec. The mass of o^2 Eri B is 0.43 $M_\odot$ – probably somewhat smaller than the mean mass of a white dwarf.

(b) *Proper Motions and Tangential Velocities.* Normally parallax programs of faint stars prefer objects with large proper motions. Search for degenerate stars independent of their proper motions has so far been very limited.

The mean annual proper motion of the 33 white dwarfs with trigonometric parallaxes is 1".33 whereas the mean for the 15 stars with luminosities estimated by Eggen and Greenstein is only 0".39. Even if we should not jump to conclusions from this difference the enlistment of some of these objects with very small proper motions

* Mitteilungen Serie A, No. 43.

in the Nearby Star Catalogue seems to be dubious: Feige 34, Feige 110, CPD $-69°$ 177, Feige 22, Case 1. The tangential velocities of these 5 objects relative to the sun are below 10 km/sec. Except Case 1, their luminosities derived from the $(M_V, U-V)$ relation of white dwarfs are fairly bright. In the group of white dwarfs with photometric distances they are responsible for an increasing mean tangential velocity (TV) with decreasing luminosity whereas the mean TV among the 32 objects with trigonometric parallaxes does not change remarkably with the absolute magnitude (Table I).

TABLE I

Luminosities and tangential velocities of white dwarfs

M_V	photom. parallax		trigon. parallax	
	n	$\overline{TV}$ km/sec	n	$\overline{TV}$ km/sec
10–11	5	7	1	44
11–12			5	65
12–13	5	26	6	38
13–14	4	39	7	43
14–15	1	103	7	70
15–16			6	73

(c) *Solar Motion.* The space velocities of 17 white dwarfs – single stars or primaries in a system – give the solar motion and velocity dispersion shown in Table II. The measured RV's have been corrected by -51 km/sec. The U-, V-, W-axes are in a

TABLE II

Solar motion and velocity dispersion (km/sec)

Class	n	$V_\odot$ (km/sec)		L_a	p.e.	B_a	p.e.	σ_U	σ_V	σ_W
D	17	20	± 4	69°	$\pm 20°$	$+21°$	$\pm 10°$	± 42	± 20	± 21
dM	202	23	1.3	71	5	$+16$	3	38	25	21
dMe	106	13	1.3	66	10	$+31$	6	31	19	13

righthand galactic system, U directed to the galactic centre. The data are compared with the results of an investigation of the early M dwarfs in the McCormick program (Gliese, 1958) which obviously represents the red dwarfs free from selection effects.

The velocity distribution of the few white dwarfs agrees fairly well with that of the dM stars. The deviations from the data of the emission-line red dwarfs seem to be larger. But in view of the large errors this result should not be over-valued.

If we do not correct the measured RV's of white dwarfs by -51 km/sec (mean Einstein effect) the dispersion among the space velocities of 16 stars is distinctly larger than the data in Table II: ± 53, ± 26, ± 30 km/sec.

(d) *Colour-Luminosity Relations*. In Figures 1, 2, and 3 the distribution of the stars in the $(M_V, B-V)$, $(M_V, U-B)$, and the $(M_V, U-V)$ plane is shown. They include only the 15 objects whose trigonometrically determined absolute magnitudes are known with probable errors not exceeding $\pm 0.^m20$. The size of the rectangles is defined by $M_V - 5 \times$ p.e., $M_V + 5 \times$ p.e. (maximal error of M_V which results from the errors in the trigonometric parallax and in the apparent magnitude); the errors in the colour data are estimated values. So we may assume that the true position of each star will be within the limits of its rectangle.

The $(M_V, B-V)$ diagram does not show much more than a real cosmic dispersion. In the $(M_V, U-B)$ plane the two groups pointed out by Eggen and Greenstein (1965) are indicated – the brighter group represented by only 4 stars. The same is true of the $(M_V, U-V)$ plane.

Obviously a star like o^2 Eri B belongs to another group than the stars of nearly the same colour but about two classes fainter. A real gap seems to be possible between both samples.

The addition of the stars with more inaccurate data does not give much more information but it obscures the answer to the question of the existence of a gap.

In the $(U-B, B-V)$ diagram of these 15 objects (Figure 4) the 4 bright degenerate stars of the upper relation (relation I of Eggen and Greenstein, 1965) are farther below

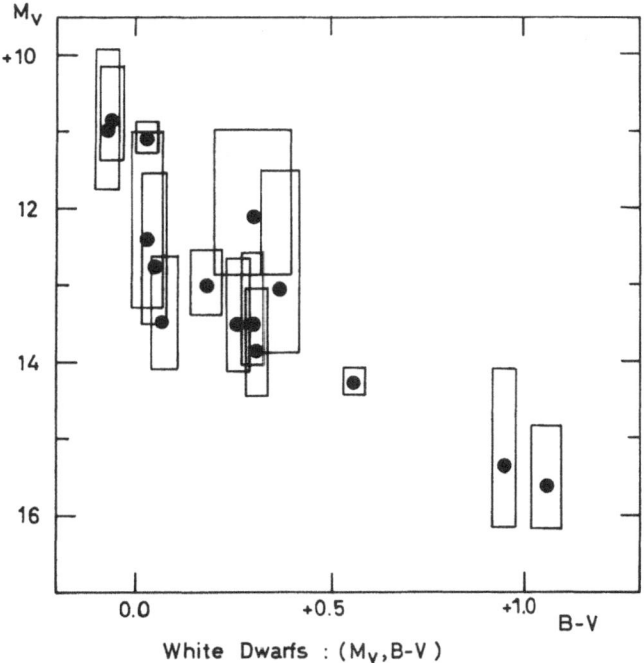

Fig. 1. The $(M_V, B-V)$ diagram for white dwarfs with trigonometrically determined absolute magnitudes, p.e. not exceeding $\pm 0^m.20$. The rectangles represent the areas whose limits indicate maximal errors (maximal error $= 5 \times$ prob. error).

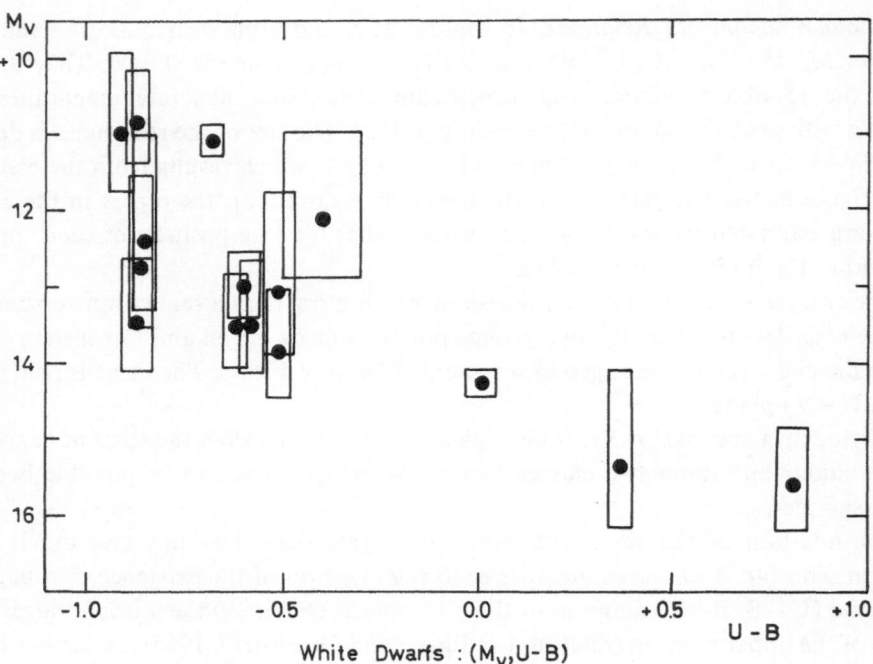

Fig. 2.   The $(M_V,\ U-B)$ diagram for white dwarfs with trigonometrically determined absolute magnitudes, p.e. not exceeding $\pm 0^m.20$. The rectangles represent the areas whose limits indicate maximal errors (maximal error $= 5 \times$ prob. error).

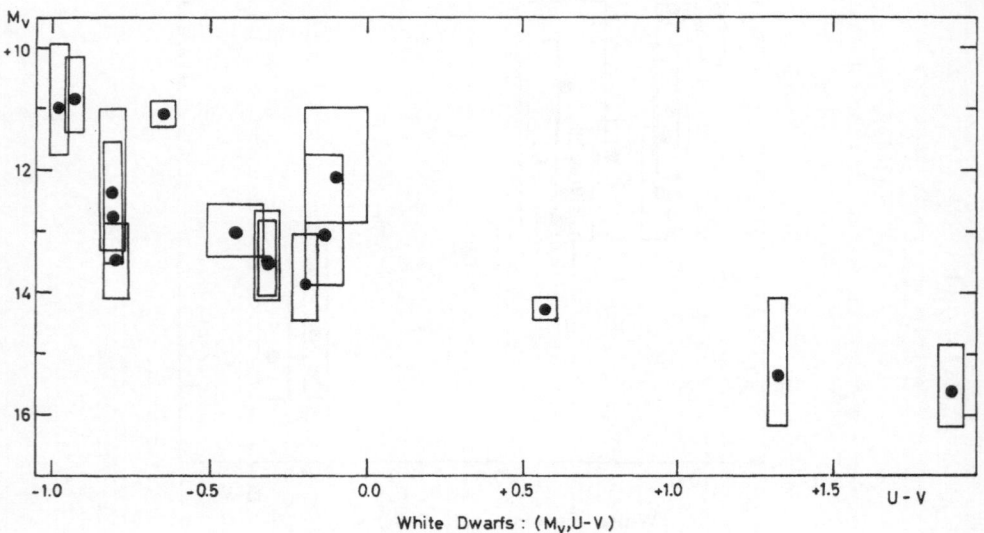

Fig. 3.   The $(M_V,\ U-V)$ diagram for white dwarfs with trigonometrically determined absolute magnitudes, p.e. not exceeding $\pm 0^m.20$. The rectangles represent the areas whose limits indicate maximal errors (maximal error $= 5 \times$ prob. error).

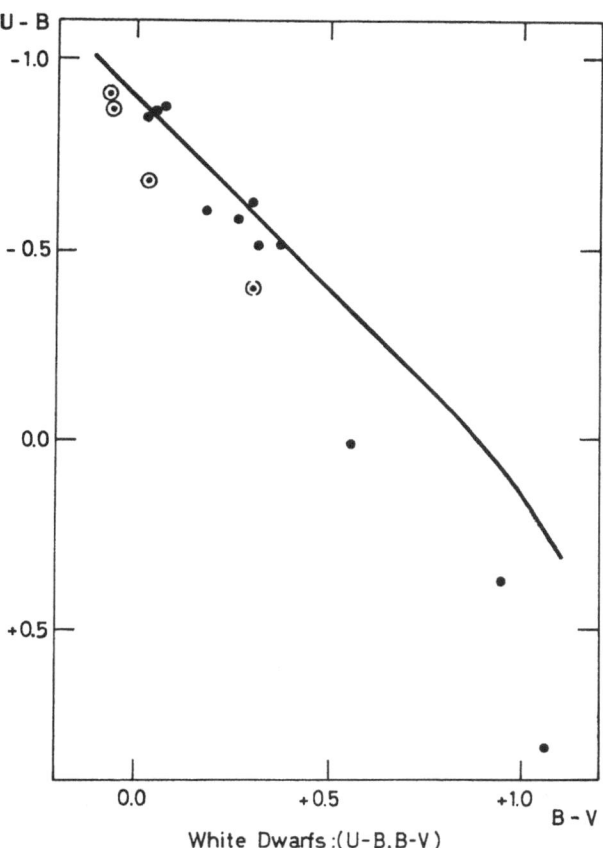

Fig. 4.   Positions of the white dwarfs with well-known luminosities in the $(U - B, B - V)$-plane. Open circles indicate the stars of bright luminosities (relation I, Eggen and Greenstein, 1965); the lowest of these 4 stars has only preliminary colour data. The continuous curve represents the black-body line.

the black-body line than the stars of the second relation. Probably there really is a slight difference between both groups of white dwarfs at least in the region of the A type stars.

(e) *Binaries and Triples.* Fourteen of the 54 nearby objects are components of double stars; two further white dwarfs belong to triples. In four systems the degenerate star is the primary. For Sirius, Procyon, and o^2 Eri the orbits and masses are known. If we estimate the masses of the other companions we may assume that in 8 or even 9 of the systems the white dwarf component will be the primary in mass.

Figure 5 shows the Hertzsprung-Russell diagram of the companions; nearly all of them are main-sequence stars, mostly red dwarfs. The 5 emission-line stars (open circles) are o^2 Eri A (K1 Ve) and C (dM5e), BD + 26°730 (dK5ep) – which probably is a member of the Hyades group –, the eclipsing binary LP 101–15 (dM3e–dM4e),

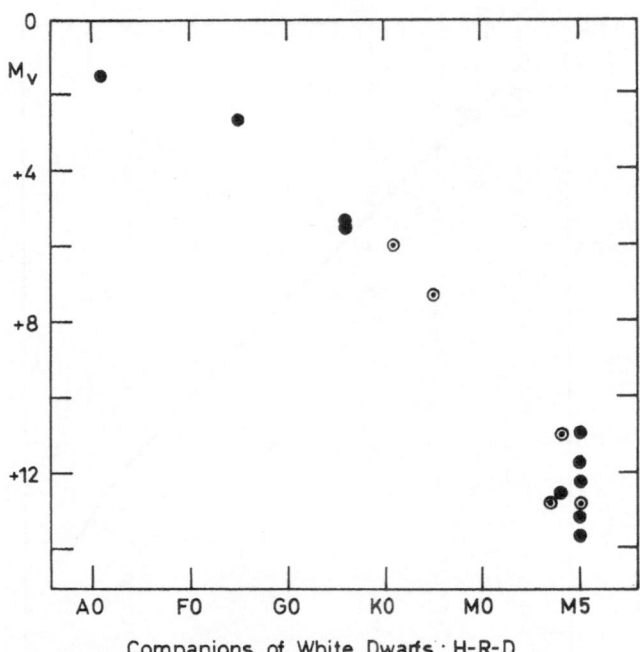

Companions of White Dwarfs : H-R-D

Fig. 5.   H-R-D of the companions of white dwarfs. Emission-line stars are shown as open circles.

and Ross 193 (dM4e). The systems +26°730/G 39–27 and Ross 193/VB 11, together
with the problem of a combination of a degenerate star with an emission-line red
dwarf, have been discussed by Eggen and Greenstein (1967). o^2 Eri and LP 101–15/16
are high velocity systems.

This small sample of binaries does not show any striking correlations. The components are from A1 V (Sirius) resp. F5 IV–V (Procyon) down to dM5 or even later
(L 745–46 B); the degenerate components are of type A, As, Awk, F, or C. The
magnitude differences "white dwarf minus companion" are from 10.5 to −3.0. The
tangential distances are from 16 AU (Procyon) to 9000 AU. But there is no correlation
between distance and type or magnitude difference.

This short report on the degenerate stars with the best known distances, luminosities,
and tangential velocities must close with the disappointing remark that these data are
far from being sufficient for detailed discussions of their problems.

References

Eggen, O. J. and Greenstein, J. L.: 1965, *Astrophys. J.* **141**, 83.
Eggen, O. J. and Greenstein, J. L.: 1967, *Astrophys. J.* **150**, 927.
Gliese, W.: 1958, *Z. Astrophys.* **45**, 293.
Gliese, W.: 1969, *Veröff. Astron. Rechen Inst. Heidelberg* **22**.
Greenstein, J. L. and Trimble, V. L.: 1967, *Astrophys. J.* **149** 283.

8. HIGH FREQUENCY STELLAR OSCILLATIONS

IV: *Photoelectric Monitoring of Southern White Dwarfs*

J. E. HESSER

Cerro Tololo Inter-American Observatory*, La Serena, Chile

and

B. M. LASKER

Cerro Tololo Inter-American Observatory*
and The University of Michigan

Abstract. Time-series data for 14 stars in the list of Eggen and Greenstein have been used to compute their power spectra, which confirm previously found quiescency in the 4 to 700 sec period range. Additionally, characteristics of the continuous power spectra are considered.

1. Introduction

Late in 1968, we undertook at the Cerro Tololo Inter-American Observatory an extension of the program begun at Princeton University to survey observationally the features of the power spectra of representative degenerate stars. While our primary motivation was to reobserve the central stars of planetary nebulae, for which the Princeton results were somewhat inconclusive, we did observe a much broader sample of about 50 stars thus including white dwarfs, central stars of planetary nebulae, old novae, U Gem stars, and other objects (see, e.g. Lasker and Hesser, 1970). The discussion for 14 stars selected from the list of white dwarfs compiled by Eggen and Greenstein (1965) is given here; those results for the planetary nebulae will be presented shortly.

2. Observations

Our data-collection and analytical techniques closely resemble those previously discribed (Hesser, Ostriker and Lawrence 1969; hereinafter referenced as HOL); only departures from those procedures are discussed further here. The observations were obtained with the 60-, 36- and No. 2 16-inch telescopes on Cerro Tololo; a standard one-channel, 1P21, offset-guiding photometer was used. The typical data set consists of UBV photometry (for identification purposes) followed by about two hours of continuous monitoring in integrated light with 2-second integrations; the UT of every seventh data point was also recorded. A 12.5″ or 27″ diaphragm was normally used, and guiding was done to better than 1″. Sky measurements were acquired at random intervals of approximately 45 min for constant and dark skies and more frequently under less favorable conditions. The white dwarf observations, viewed in

---

* Operated by the Association of Universities for Research in Astronomy, Inc., under contract with the National Science Foundation.

the context of our program, were generally intended to serve as standards of quiet objects and multiple observations of the same star were seldom obtained. Furthermore, such stars were more often observed when, for some reason, conditions were suspect. Despite such reasons for expecting that the observations of white dwarfs reported here will not reach the ultimate limit of which the techniques are capable, we have attained, on the average, an improved limit on the 'noisiness' of these stars.

The University of Michigan's data system, being extremely compact, was transported to Chile by one of us (B.M.L.) for this program, and was used to register all data. This apparatus consists principally of Solid State Radiation pulse amplifiers, LeCroy 50 MHz scalers, and various micrologic control circuits. The scaler is gated by a precision crystal-controlled timer. In the monitoring mode, this system repeatedly makes an integration, loads the count into a punch buffer in 90 μsec, and begins another integration cycle while punching the previous count. The 30 frame/sec speed of the paper-tape punch permitted outputting 0.1 sec integrations with no data loss, but normally 2 sec integrations were used to minimize tape consumption. For facility in analysis, all paper tape was converted to magnetic tape at the end of the observing run. The observations, together with certain analytical parameters discussed below, are summarized in Table I, the format of which is nearly identical to that used in HOL.

3. Analysis and Results

Our computations of the power spectrum for each data-set of Table I were performed by the standard techniques (see, e.g. Blackman and Tukey (1958), Bingham *et al.* (1967), and HOL). For all spectra a window nearly the total width of the data-set was used, and calculations were made for all frequencies from the Nyquist frequency to the reciprocal length of the data set. The resultant spectra were examined for evidence of harmonic activity in the original data sets; this, of course, would manifest itself as a delta-like function in the power spectrum. The amplitude of the largest local maximum, A_{max}, and its associated period, $T(A_{max})$, are given in Table I for each data set. With the exception of G44-32 (EG 72), our results are consistent with those of HOL: white dwarfs are quiescent in the period range from 2 to 700 sec. Furthermore, we note that the generally superior sky conditions at Cerro Tololo enable us to reduce the mean values of the noise statistic, Q (see HOL), and A_{max} from the Princeton values by a factor of about 2 (see Table II). G44-32, whose recently discovered variability has been discussed elsewhere (Lasker and Hesser, 1969, Warner *et al.*, 1970), is seen to stand apart from the other stars in Table I in all three parameters: Q, A_{max}, and x, the slope of the low frequency portion of the power spectrum (see below).

In addition to examining the local behaviour of power spectra for evidence of harmonic activity, inspection of the overall properties of the spectra for continuous trends is of some importance. Here, for example, is where the effects of flickering would appear. From log-log plots of $P(f)$ it is clear that the power spectra generally consist of two parts: a flat component due to the photon statistics of observing, and a component decaying with frequency due to various aperiodic fluctuations in the

Summary of white dwarf observations †

Designation Discover	EG	Mag.	Sp. T.	Date	Time (UT)	Tel.	N	A_{max} (mag)	$T(A_{max};$ sec)	Q	x
LB3303	21a	11.2	(DA)	01/05–06/69	02:20:02	36	3627	0.0008	6.3	0.004	−0.01
L587-77A	22*	13.9	DAs	01/06–07/69	01:17:11	36	3681	0.009	455	0.038	−1.47
θ² Eri B, C	33*	9.52	DA	12/31/68–01/01/69	02:30:35	36	3695	≥ 0.008	120	0.004	−0.04
L879-14	41	14.10	4670	01/19–20/69	01:45:04	16	4051	0.016	178	0.029	−0.66
L1244-26	46	13.40	DA	01/08–09/69	04:34:28	36	2715	0.003	482	0.006	−0.81
L745-46A	54	12.98	DF	12/28–29/69	04:38:06	36	1147	*	*	*	*
L97-12	56	14.5	DC	01/30–31/69	06:35:00	36	3140	*	*	*	*
L532-81	62	12.0	DA, Fs	12/30–31/68	04:49:14	36	5461	0.004	607	0.008	−0.76
				12/31/68–01/01/69	05:38:57	36	3975	0.0008	293	0.005	−0.18
				02/05–06/69	04:22:58	16	2729	0.002	5.3	0.008	−0.29
LDS275 AB	66*	15.0	DC	01/22–23/69	04:34:00	36	3175	0.002	67.7	0.011	−0.43
L825-14	70*	12.97	DAn	01/15–16/69	05:12:23	16	3633	0.003	57.7	0.013	−0.24
G44-32	72*	16.55		01/24–25/69	07:34:06	36	2071	0.016	14.0	0.026	−0.15
				02/20–21/69	05:15:29	60	10795*	0.034	862	0.029	−1.16
				02/21–22/69	06:53:18	60	4916*	0.036	315	0.023	−1.42
				04/11–12/69	04:48:12	50*	5041	0.035	420	0.016	−1.04
				04/12–13/69	04:11:28	50*	6706	0.023	630	0.026	−1.39
F46	81*	13.24	DO-B	01/21–22/69	06:23:36	16	4137	0.005	6.1	0.023	+0.01
L145-141	82	11.44	4670	01/06–07/69	06:58:13	36	2365	0.0009	315	0.004	−0.31
L845-70	122	14.30	DC?	04/12–13/69	09:12:25	50*	~7200	*	*	*	*

Notes:

† Where an asterisk (*) appears next to the EG number or in the body of the table, a note pertaining to that star follows. Unless otherwise noted below, data were acquired using no filter and 2 sec integration times.

EG 22. Both stars included in diaphragm; second star is sdM3, V = 15.6, B − V = +1.7. Noisiness of record due to guiding difficulties and Moonlight interference. Q, A_{max}, and x-values not used in forming means.

EG 33. Both stars B and C were in diaphragm; V filter used.

EG 54. Automatic features of editing program failed due to shortness of run; inspection of analog record indicates that the object is non-variable in the higher frequency range.

EG 56. B filter used with an integration time of 1.060 sec. Error in computation of power spectrum occurred which does not allow us to compute A_{max} or x from present spectrum.

EG 66. Double white dwarf system; both stars taken with cautious guiding to keep third star out of diaphragm.

EG 70. Dramatic change in seeing occurred during run, as well as several guide errors.

EG 72. None of the derived parameters for this star are included in the means discussed in text. The February data were acquired with 1 sec integration times. The April data were obtained with the Kitt Peak National Observatory 50-inch telescope.

EG 81. U filter used to minimize effect of faint companion star. Eggen and Greenstein note that F 46 may be a sdO star.

EG 122. Analysis of automatically recorded data have not been performed to date; inspection of analog record indicates that the object is non-variable and it is included here for completeness. KPNO 50-inch telescope. Possibly a λ 4670 star.

TABLE II

	Princeton				Cerro Tololo			
	No. Obs.	M	σ_i	σ_m	No. Obs.	M	σ_i	σ_m
Q	42	0.019	0.016	0.002	12	0.011	0.008	0.002
A_{max}	38	0.012	0.009	0.001	12	0.004	0.004	0.001

Notes: M is the mean, σ_i is the standard deviation of an individual datum and σ_m is the standard deviation of the mean (Poisson statistics assumed).

light (see, e.g., Figure 1). As the individual log-log plots contain too much information of dubious significance to justify publication, we have chosen to characterize the decaying part of the power spectrum by the parameter x, where $P(f) \alpha f^x$. The results for x given in Table I were obtained by a least squares fit for periods from

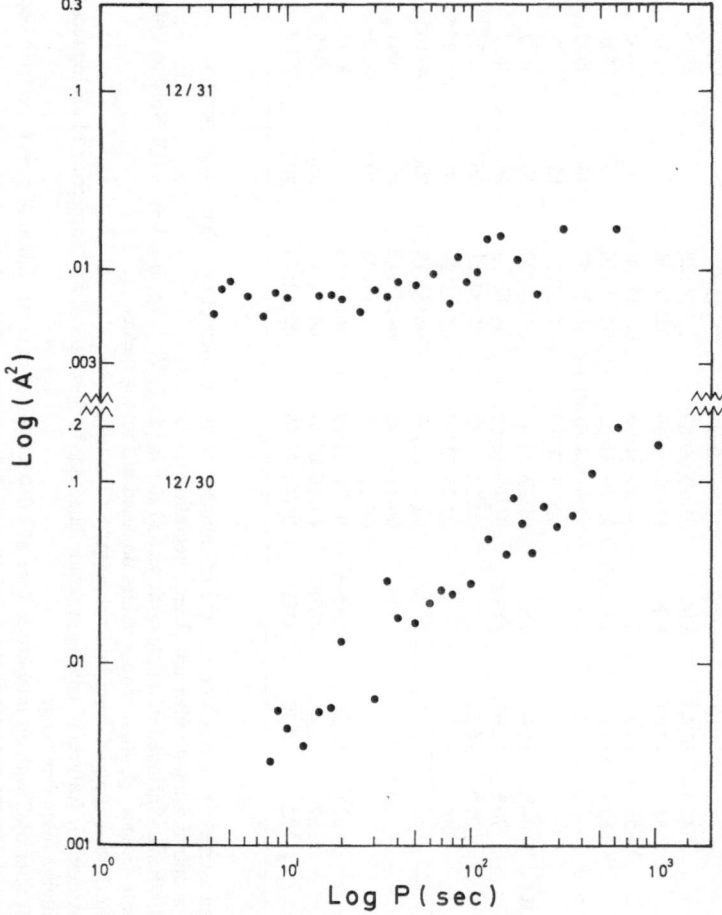

Fig. 1. Smoothed spectra computed from high resolution spectra of L532-81 (EG 62).
Ordinate is amplitude squared in relative units normalized such that unity is the approximate amplitude of the 1 % calibration peak introduced in the analysis and corrected for sky contributions.

$T/10$ to $T/83$, where T is the record length. The mean value of x for observations of 11 stars is -0.34 ± 0.09 (standard deviation). Unfortunately our data were not secured, nor have they been analyzed, in a fashion that lends greatest reliability to the continuum behaviour at the lowest frequencies, in that point-to-point sky subtraction has not been made experimentally (for a description of the analytical formalism used see HOL). Furthermore, a cursory examination of the astronomical literature on seeing and scintillation (Mikesell, 1955; Stock and Keller, 1960; Rozhnova, 1966; and references cited therein) does not provide a ready comparison from either observation or theory. It will nonetheless be interesting to compare the mean white dwarf result with those obtained from other classes of objects, where the data were acquired and analyzed under similar conditions; we shall report upon that comparison in future papers.

Acknowledgement

It is a pleasure to acknowledge the National Science Foundation for their support of the Cerro Tololo Inter-American Observatory and the University of Michigan for providing funds for the construction and transportation of the data system to Chile. The computing facilities of the University of Michigan and of the Associated Universities for Research in Astronomy were used very extensively in this program.

References

Bingham, C., Godfrey, M. D., and Tukey, J. W.: 1967, *IEEE Trans. Audio and Electroacoustics* **15**, 56.
Blackman, R. B. and Tukey, J. W.: 1968, *Measurement of Power Spectra*, Dover Publications, New York.
Eggen, O. J. and Greenstein, J. L.: 1965, *Astrophys. J.* **141**, 83.
Hesser, J. E., Ostriker, J. P., and Lawrence, G. M.: 1969, *Astrophys. J.* **155**, 919 (HOL).
Lasker, B. M. and Hesser, J. E.: 1969, *Astrophys. J.* **158**, L171.
Lasker, B. M. and Hesser, J. E.: 1970, *IAU Info. Bull. Var. Stars*, No. 415.
Mikesell, A. H.: 1955, *Publ. U.S. Naval Obs., 2nd ser.* **17**, 139.
Rozhnova, I. P.: 1966, in *Optical Instability of the Earth's Atmosphere* (ed. by N. I. Kucherov), Israel Program for Scientific Translations, Jerusalem, p. 139.
Smak, J.: 1967, *Acta Astron.* **17**, 255.
Stock, J. and Keller, G.: 1960, in *Telescopes* (ed. by G. P. Kuiper and B. M. Middlehurst), University of Chicago Press, Chicago, p. 138.
Warner, B., van Citters, G. W., and Nather, R. E.: 1970, *Nature* **226**, 68.

9. NEW SPECTROSCOPIC RESULTS ON
SUBLUMINOUS STARS, V

J. L. GREENSTEIN

Hale Observatories,
California Institute of Technology,
Carnegie Institution of Washington

Abstract. Determination of temperature and surface gravity by colors and hydrogen-line profiles have been carried out for hot halo stars. A narrow horizontal branch is found stretching to above 40000 K; the hot O subdwarfs show a nearly vertical sequence, dropping towards the hot white dwarfs.

Spectra for 285 white dwarf stars have been obtained, and the classification scheme is reviewed. Theoretical problems of these spectra remain, largely, unsolved.

The red subluminous stars found by Eggen were studied spectroscopically; among 68 stars only one new red degenerate star was found. The others are very metal-poor, high-velocity stars with large ultraviolet excess.

1. Introduction

I will discuss three separate topics (A) the approach to the white-dwarf stage among the hot-horizontal branch and subdwarf B and O stars, (B) recent data on normal and peculiar white dwarfs, (C) theoretical problems, (D) the suspected red subluminous stars.

A. THE SPECTROSCOPIC EVIDENCE ON THE PRE-WHITE DWARF STARS

The discovery of interesting varieties of stellar spectra with a goal of determining composition has been successful, among ordinary stars. Interesting composition differences and the effects of nucleosynthesis are thus observed. For the degenerate stars we have not been so optimistic, since pressure broadening, gravitational separation, and unknown opacity sources present as yet unsolved problems when we attempt to deduce the true composition of the star. Since some hot degenerate stars show effects of nucleosynthesis in the form of low-hydrogen content, high-helium and high-carbon content, some core material has been brought to the surface; other cool degenerate stars show weak or no lines i.e., have low-surface metal content. In the pre-white dwarf stage, it was suggested in 1960 that the surface He/H ratio was low; later observations confirmed this, for halo horizontal-branch stars. My recent work on halo blue stars was presented as the Henry Norris Russell Lecture of the American Astronomical Society, and will be published in detail later. The major relevant features are: (1) The sample of halo stars were selected by blue color only, in the galactic polar caps, not by proper motion. (2) The UBV colors determine effective temperature, T_e, moderately well from 12000 K up to 30000 or 35000 K, independent of the surface gravity g. (3) The model atmospheres available now predict hydrogen-line profiles as a function of (T_e, g) in the desired range. My spectra (from 18 to 190 Å mm^{-1}) yield profiles of sufficient accuracy to determine g, given T_e.

I selected 170 stars from 9th to 16th mag, with negative B−V colors; the brighter

Luyten (ed.), White Dwarfs, 46–60. All Rights Reserved.

stars are largely uninteresting, in that they seem normal, even though far from the galactic plane. Of course, their existence does present a serious evolutionary question. The fainter stars were 'abnormal' in that they belonged to groups identified by visual spectral classification as (1) 'horizontal branch' (HBA, HBB) recognized by sharp, numerous hydrogen lines i.e., low g and weak or no metallic lines; (2) subdwarfs (sdB, sdO) with broader shallower H lines, distinguished by very weak or no He I in the sdB, and very shallow H lines in sdO, which might be accompanied by He II lines; (3) white dwarfs with very broad hydrogen lines (DA), helium lines (DB), or no detectable lines (DC). A few peculiar spectra will be discussed later. The visual classification of some of these stars was published in Greenstein (1966) and some luminosity estimates in Greenstein and Eggen (1966). Quantitative analyses by use of UBV (or other) colors and Hγ profiles have been carried out by Searle and Rodgers (1966), Sargent and Searle (1968) and Newell (1969). In this investigation fainter stars were studied and often at higher dispersion. A few general remarks about the new results may be interesting. The 'normal' stars amounted to 20%, but had a bright mean apparent magnitude 10.4. The HBA and HBB stars accounted for 22%; the sdB 16%; the sdO 13%, and were faint, $\langle m = 14.1 \rangle$. Twenty-two stars had composite colors and spectra (i.e., unresolved binaries), showing that the percentage of binaries is not very low in the halo. Fifteen percent of the halo stars were white dwarfs found by color i.e., not selected for motion. Selection effects do exist, in that I observed the bluest stars first, and a very large number of HBA stars near $B - V = 0.0$ were not observed. Thus, the fraction of very hot stars is too large. The frequencies per volume of space for the different groups are obviously quite different, and the 'normal' stars suffer from the density decrease with z-coordinate. A preliminary solution indicates a luminosity function such that the true number per unit volume increases as exp $(0.65 \, M_V)$ from $M_v = -4$ to $+5$. This will be recomputed on a M_b scale.

The value of g found depends on the assumed mass; since all the stars may be assumed to be highly evolved, it seems best to adopt a constant mass. Some sdO stars may become planetary nebulae, to lose excess mass, and a few have associated planetaries of low-surface brightness. The mass assumed in Table I was 1 $m_\odot$ in the relation

$$M_b = 2.5 \log g + 10 \log \theta - 2.5 \log (m/m_\odot) - 5.82. \tag{1}$$

Therefore M_b should be increased by $+0.75$ if $m = 0.5 \, m_\odot$. The accidental errors are of the same order, except for the very hot stars. Since the colors have errors ± 0.03, the value of the reddening-free color parameter Q has errors ± 0.042,

$$Q = (U - B) - 0.72 (B - V), \tag{2}$$

and that of $\theta \pm 0.021$. Then at $\theta = 0.12$, the highest temperature used, M_b has errors ± 0.9 mag. In addition, the g from line profiles have errors, since the line strength depends on both θ and g.

Table I provides temperatures and luminosities determined spectroscopically for a selection of three types of objects from our larger catalog of results. The bolometric luminosities are plotted in Figure 1. The cooler horizontal-branch stars, near $B - V = 0.0$ have colors which may depend on g, so we show only stars with $T > 10000$ K.

TABLE I

Sample data for hot horizontal-branch and subdwarfs

FB No.	Name	V	B − V	U − B	θ	log g	M_b	He I λ4388	He I λ4471	He II λ4686
					Subdwarf O					
3	TS 144	12.9	− 0.17	− 1.18	< 0.12	5.0:	< − 2.5	1.14	2.46	2.77
13	TS 201	13.2	− 0.24	− 1.04	0.17	5.8	+ 1.0	< 0.12	0.30	0.74
26	HZ 3	12.9	− 0.14	− 1.10	0.12	< 6.0:	< 0.0	0.96	1.94	2.52
42	Abell 31	15.5	− 0.31	− 1.28	< 0.12	> 5.5	− 1.2	< 0.20	< 0.20	1.17
45	GD 299	12.1	− 0.28	− 1.15	0.14	5.4:	− 0.9	0.37	1.20	1.93
48	GD 300	12.8	− 0.33	− 1.19	0.14	5.2:	− 1.4	0.29	1.38	2.22
51	F 34	11.2	− 0.30	− 1.35	< 0.12	5.9	− 0.3	< 0.03	< 0.03	2.47
62	F 46	13.2	− 0.25	− 1.16	0.13	6.9:	+ 2.6	1.81	3.09	1.28
89	F 67	11.8	− 0.33	− 1.21	0.13	5.2:	− 1.7	< 0.12	< 0.12	1.31
102	HZ 38	14.2	− 0.27	− 1.16	0.13	7.2	+ 3.3	1.67	3.13	1.17
115	HZ 44	11.7	− 0.29	− 1.19	0.13	5.7	− 0.4	1.68	2.99	2.32
167	F 110	12.5	− 0.30	− 1.20	0.13	6.5	+ 1.6	0.24	0.43	1.43
					Subdwarf B					
1	TS 135	13.2	− 0.18	− 0.80	0.25	5.0	+ 0.6	< 0.12	< 0.12	< 0.12
10	TS 183	12.6	− 0.24	− 1.05	0.17	5.5	+ 0.2	< 0.09	< 0.09	< 0.09
11	F 11	12.1	− 0.26	− 1.02	0.19	5.3	+ 0.2	< 0.12	p	< 0.20
47	GD 104	15.9	− 0.35	− 1.23	0.13	5.6	− 0.4	< 0.20	< 0.20	< 0.20
49	GD 108	13.6	− 0.21	− 0.91	0.22	5.3	+ 0.8	< 0.05	< 0.05	< 0.05
56	F 38	13.0	− 0.23	− 1.00	0.18	5.6	+ 0.7	< 0.20	0.39:	< 0.20
66	F 55	13.6	− 0.38	− 1.25	0.13	5.9:	+ 0.1	< 0.12	< 0.12	< 0.12
67	HZ 17	15.5	− 0.22	− 1.00	0.18	5.5	+ 0.5	< 0.20	< 0.20	< 0.20
69	HZ 19	15.6	− 0.22	− 1.09	0.15	> 6.6:	> + 2.4	0.20:	0.88	< 0.20
84	F 65	12.0	− 0.23	− 0.99	0.19	5.1	− 0.3	< 0.03	0.46	< 0.03
86	F 66	10.5	− 0.32	− 1.06	0.19	5.3	+ 0.2	0.20	1.06	< 0.03
106	HZ 39	15.4	− 0.32	− 1.16	0.14	6.5:	+ 1.9	< 0.20	< 0.20	< 0.20
109	HZ 40	14.6	− 0.24	− 1.02	0.18	5.6	+ 0.7	< 0.20	0.47	< 0.20
110	F 75	14.5	− 0.21	− 0.93	0.21	5.0	− 0.1	< 0.12	< 0.12	< 0.12
117	F 81	13.5	− 0.22	− 1.02	0.18	5.6	+ 0.7	< 0.20	0.45	< 0.20
127	F 91	13.4	− 0.28	− 1.07	0.17	5.5	+ 0.2	< 0.20	0.55	< 0.20
143	TN 788	13.2	− 0.18	− 1.00	0.17	5.5	+ 0.2	0.35	1.20	< 0.12
146	TN 245	13.9	− 0.24	− 0.90	0.23	5.0	+ 0.3	< 0.12	< 0.12	< 0.12
165	F 108	12.2	− 0.28	− 1.06	0.18	5.1	− 0.5	< 0.12	< 0.12	< 0.12
					Hot Horizontal Branch B − V ≤ − 0.05					
20	F 23	11.9	− 0.11	− 0.44	0.38	4.0	+ 0.7	< 0.20	< 0.20	< 0.20
50	GD 113	11.6	− 0.09	− 0.49	0.34	3.8	− 0.3	< 0.05	< 0.05	< 0.05
52	+ 40°449-34	11.3	− 0.12	− 0.48	0.35	3.7	− 0.4	0.29	0.67	–
63	F 48	13.3	− 0.17	− 0.68	0.29	3.8	− 1.0	< 0.20	< 0.20	< 0.20
70	+ 36°2242	10.0	− 0.08	− 0.26	0.43	3.5	0.0	< 0.05	< 0.05	< 0.05
75	F 59	11.6	− 0.06	− 0.10	0.49	3.1	− 0.5	< 0.05	< 0.05	–
78	HZ 24	11.9	− 0.05	− 0.15	0.47	3.7	+ 0.9	< 0.05	< 0.05	< 0.05
79	+ 49°2137	10.7	− 0.14	− 0.60	0.31	3.8	− 0.7	p	0.25	< 0.03
82	HZ 26	13.8	− 0.06	− 0.30	0.41	3.6	0.0	< 0.20	< 0.20	< 0.20
85	HZ 30	13.6	− 0.14	− 0.62	0.31	3.7	− 1.0	< 0.20	0.33	< 0.20
103	HZ 47	15.3	− 0.13	− 0.74	0.26	4.6	+ 0.5	–	0.50	< 0.20
105	HZ 37	12.5	− 0.08	− 0.18	0.46	3.4	0.0	< 0.05	< 0.05	–
107	F 74	12.0	− 0.10	− 0.32	0.41	3.0	− 1.5	< 0.09	0.40:	< 0.09

Table I (continued)

FB No.	Name	V	B − V	U − B	θ	log g	M_b	W(Å) He I λ4388	He I λ4471	He II λ4686
					Hot Horizontal Branch B − V ≤ − 0.05			He I λ4388	He I λ4471	He II λ4686
108	HZ 42	14.6	− 0.14	− 0.54	0.33	3.8	− 0.4	< 0.12	< 0.12	< 0.12
112	F 76	15.2	− 0.12	− 0.65	0.29	4.0	− 0.5	< 0.20	< 0.20	< 0.20
116	HZ 45	12.8	− 0.15	− 0.57	0.33	3.8	− 0.4	< 0.05	< 0.05	−
121	F 86	10.1	− 0.15	− 0.66	0.30	3.5	− 1.6	0.15:	0.50	< 0.05
128	F 92	11.5	− 0.13	− 0.61	0.31	3.8	− 0.7	p	0.50:	−
160	PHL 25	12.0	− 0.14	− 0.68	0.28	4.4	+ 0.3	0.93	1.25	< 0.12

(A large number of apparently normal, main-sequence stars are not plotted.) Two globular cluster B stars are shown, with M_V as directly determined and with bolometric corrections from Mihalas (1965). Considering the errors of observation, the H-R diagram is surprisingly clean, showing a horizontal branch centered near $M_b = 0$, from 10000 to 50000 K, with a very small scatter. The branch is continuous with that found for $T < 10000$ K, and through the RR Lyrae variables. No present theoretical model of horizontal-branch evolution can explain this long and narrow track. The observed major difference from globular clusters is that the branch in clusters terminates at

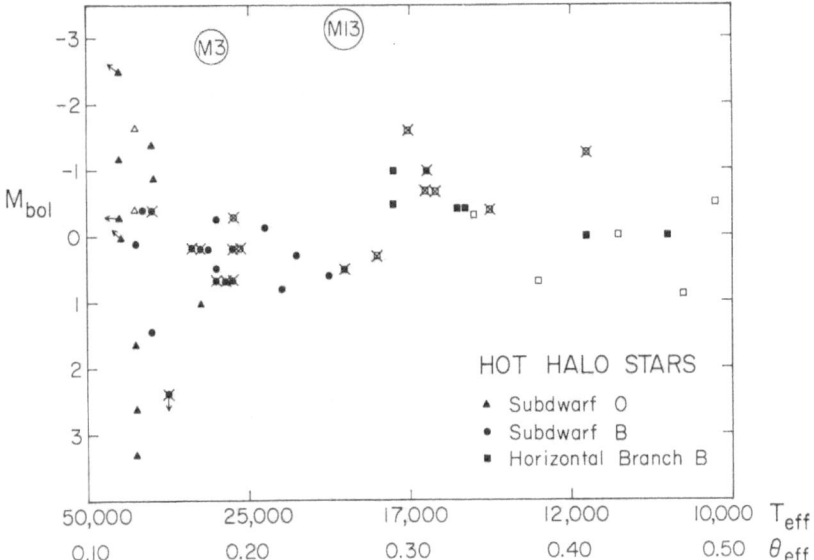

Fig. 1. Spectroscopically deduced bolometric magnitudes and temperatures of the hot halo stars. Open symbols are bright nearby stars ($m < 12$). The symbol M 3 represents the globular cluster O star in M 3, and M 13 the B star in M 13. Arrows indicate direction of possible errors. All sdO stars (triangles) have He II and sometimes He I; sdB (circles) and HBB (squares) do not show He I unless an × is added.

about $T \leqslant 17000$ K. The halo stars do not all have the same age and mass, as do stars in a cluster. Above 30000 K the present H − R diagram shows an abrupt change, with larger scatter caused by errors of θ and g. But the spread in M_b is almost certainly real; some stars in Table I are nearly white dwarfs spectroscopically. It is known that hotter and brighter subdwarfs are found as nuclei of classical planetary nebulae, and fainter ones in old novae and other binaries. Since my temperature discrimination is poor at small θ, it is possible that some sdO stars are evolving upwards and to the left, while others are dropping vertically downwards towards the white dwarf. The sdO star in M 3 (Strom, 1970) and the B star in M 13 (Stoeckly and Greenstein, 1968) lie well above and to the right of these stars, and represent a different and probably earlier stage of evolution.

B. THE SPECTRA OF THE WHITE DWARFS

A series of 3 papers with Eggen (EG I-II-III) a fourth (GR IV Greenstein, 1969a) and this report give spectra of approximately 265 white dwarfs for most of which colors and motions are known. New data on 18 more have been published (GR VI Greenstein, 1970) with added observations of EG stars (numbers less than EG 266), using a series number labelled GR, starting with EG 267. For some of these photoelectric data is missing; since only a few binaries are included, no attempt is made to give absolute magnitudes and space motions as was done before. My classification system for 285 white dwarfs is described below and has been used previously; white dwarfs differ in temperature, surface gravity and composition and therefore require a complex classification scheme. The UBV colors are not sufficient to determine the spectrum, although they correlate with dominant features over certain ranges. J. Graham (unpublished) has made considerable progress with Strömgren-type color systems. The following paragraphs attempt to describe the major type of white dwarfs as now recognizable at about 200 Å mm^{-1}.

DA: Hydrogen lines of various strengths and half widths are seen, with a maximum equivalent width $(W = 40$ Å) at $U - V = -0.55$ mag, and a width at half-central absorption $w_{0.5} = 55$ Å. Some DA stars have weak H lines for their color, are called DAwk, but DAwk occurs at both color extremes $(U - V = -1.6$ and -0.2 mag) of the DA group. Sharp-lined stars DAs appear, and especially for $U - V > -0.3$, dominate. Line blanketing and new opacity sources affect the models. Balmer lines perturb the B − V colors. In none are lines detectable beyond H 10 or H 11. Near $U - V = -0.2$ lines become very sharp and weak, but the Stark or pressure broadening persists, since the cooler DAs stars do not show even H 8, although Hγ is sharp. A new temperature scale for DA is being prepared by Oke. Most lie below the blackbody line in UBV.

DB: Over a limited range of color, near $U - V = -1.0$, a considerable fraction show only lines of HeI, without any trace of HeII or H. The $W(H\gamma)$ is reduced by a factor of at least five to ten. Theoretical analysis indicates very high He/H ratio; they lie near the blackbody line.

DO: A few very blue, hot stars (50000 to 100000 K) show HeII, sometimes HeI, and weakly H. These have high bolometric luminosity, but are below the hot sdO stars of Figure 1. Note that at 57000 and 0.01 $R_\odot$, we have $L = L_\odot$ i.e., $M_b = +4.7$. Oke believes that EG 86 (HZ 21) has $T_{eff} = 100000$. DO, DAwk, hot sdO, all lie close together and near the blackbody line.

DC: Many stars are found with no detectable lines; in the best observed cases, the central absorption <0.05 i.e., no line has $W > 1.2$ Å. These are found with $U - V$ colors from -0.7 to $+1.8$; however, at a given $B - V$, DC stars may have different $U - B$ colors. Therefore, some may have appreciably different line absorptions e.g., blended, broad invisible lines.

DA, F; *DG, DK*: So few cooler degenerate stars are bright enough for detailed study that only a rough classification exists based on metallic lines. The first appearance of Ca II in stars with sharp H lines marks DA, F. The DG stars have Ca II, and a few metallic lines e.g., Fe I and Mg I blends in the ultraviolet. DK stars are redder and have

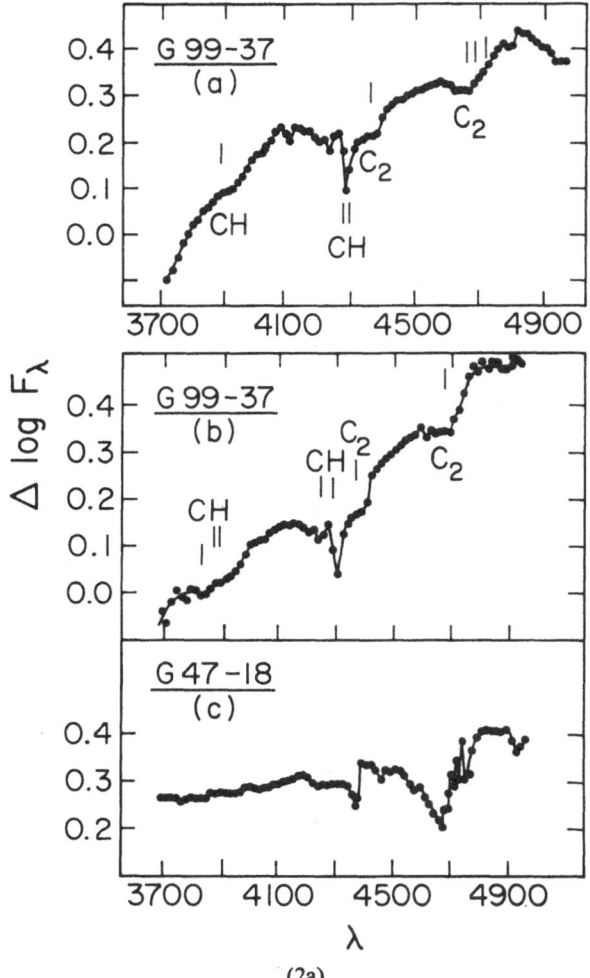

(2a)

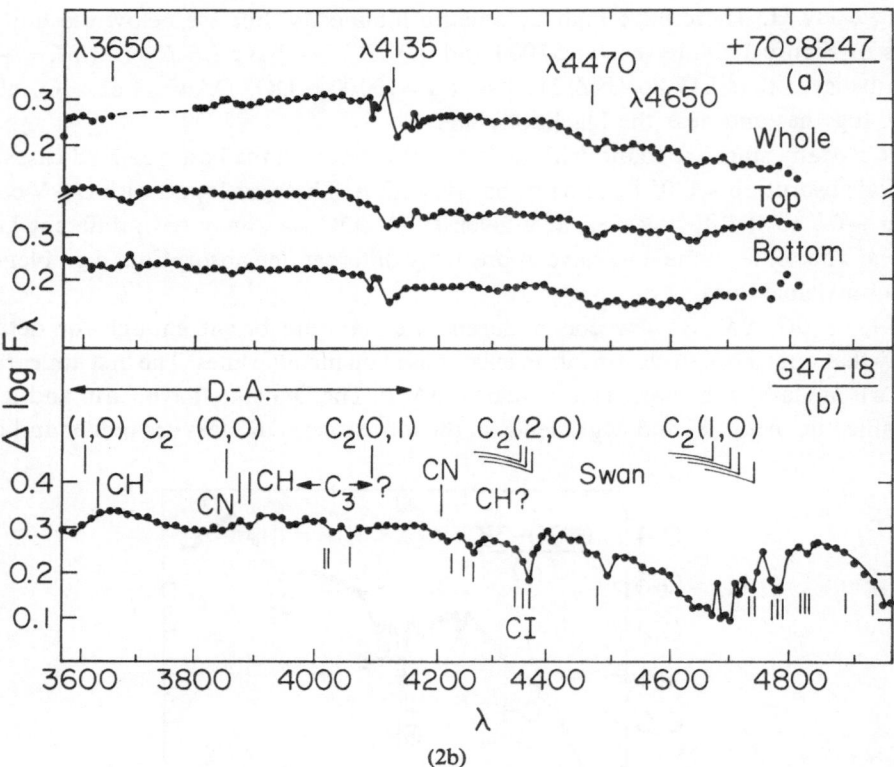

(2b)

Fig. 2a.   Intensity tracings (unsmoothed) of peculiar white dwarfs; $\Delta \log F_\lambda$ is given with respect to a DC star EG 78. Broad bands of molecular carbon, C_2, and CH are seen (a, b) in G99-37 (EG248); C_2 and C_I are seen in G47-18 (EG 182) (c), and at higher dispersion (b), in Figure 2b. The $\lambda 4135$ band dominates 3 tracings of different parts of the same 90 Å mm^{-1} plate of $+70°8247$ (EG 129) (a) in Figure 2b. Weak bands at $\lambda\lambda 3660, 4480, 4650$ are also present.

Ca II and Ca I (weak). Metallic lines vary greatly, appearing first at $U-V=-0.5$, and still appear in the reddest, which have $U-V=+1.9$.

     DM: All candidates for DM are faint (17th mag photographic). The lines seen are Ca II, Ca I, a broad depression at $\lambda > 4227$ which may be Ca$_2$ (quasi-molecule, suggested in late dM stars) or a blend of metallic lines, and blurred weak bands of TiO in the green. The properties of metal-poor sdM stars are poorly known. They have weak TiO, but usually strong MgH, and very sharp metallic lines; at adequate resolution they should be distinguishable from DM. The colors of DM stars must be affected by blurred atomic or molecular line blanketing. The colors show changes of 0.5 to 0.8 mag in $U-B$ at a fixed $B-V$ from near the main sequence to above the blackbody line.

     λ4670, λ4670p: The wavelength of the shallow-broad feature detected for these carbon-rich stars is shortward of the (1.0) head of C_2. The (0, 0) head was also seen, and is strong. In one hot carbon star, EG 182 (G47-18) the C I lines are also strong.

One single, and very puzzling example, EG 248 (G99-37) has CH strong and blurred, and strong C_2 bands as well.

$\lambda 4135$: One star of this type is known, EG129, (Greenwich $+70°8247$). This most exciting object has a band-like feature, deepest at $\lambda 4135$, degraded to the red, and weak features at $\lambda\lambda 4475, 3910, 3650$ (Greenstein and Matthews, 1957). It has just been established by Kemp and Kemp *et al.* (1969, 1970) that this star shows several percent of circular polarization and must have a magnetic field near 10^7 G, (Figure 3).

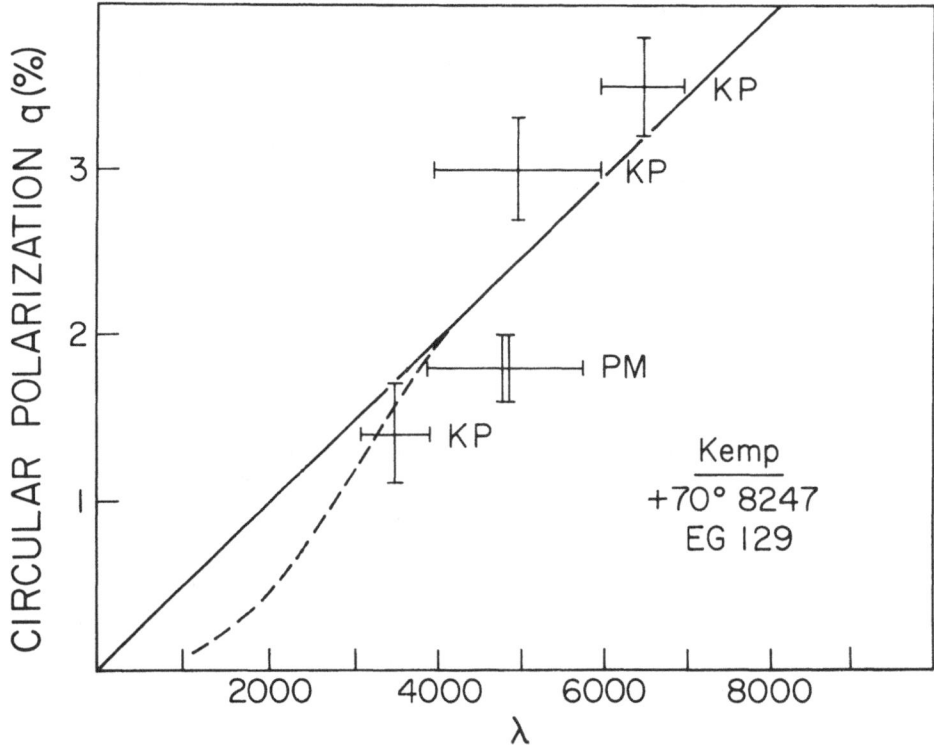

Fig. 3. Circular polarization in EG 129, $+70°8247$, as a function of wavelength (from Kemp *et al.*, 1970b).

C. NEW THEORETICAL ASPECTS ON WHITE DWARF SPECTRA

The explanation of the spectroscopic features by model atmospheres is most advanced for the DA stars. Here, Terashita and Matsushima (1969) and currently, D. Peterson, H. Shipman, and J. B. Oke (at Caltech) have constructed model atmospheres with all known opacity sources. The models have gradually improved in their ability to represent both the continuum flux and the Balmer lines. Line blanketing has severe effects on the ultraviolet continuum, and must be included. Since metals may have abnormally low abundance, cool models involve some uncertainty; convection may appear and alter the temperature gradient. The helium stars have lower opacity (Weidemann,

Bues) and a very high He/H ratio is required in DB stars to prevent H lines from appearing. However, the broadening theory is much less complete for He I lines, the He I continua are complex and non-hydrogenic, and the helium-star models need much further study. The possibility that very rapid rotation stabilizes a star of mass greater than Chandrasekhar's limit was explored by Ostriker and Bodenheimer (1968) and the effect on the DA spectrum recently studied by Wickramasinghe and Stritt-matter (1970) (following a suggestion by Sobolev). They find that rapid rotation cannot wipe out the Balmer lines and produce a DC white dwarf. Gravity darkening at the limb reduces the flux from the high-velocity equatorial regions.

Observations of details of the continuum spectrum, and interpretation by model atmospheres, together with H-line profiles should permit establishment of a white-dwarf temperature scale, and determination of surface gravity. At present the single-channel spectrophotometric scans are being improved on by use of a multi-channel spectrometer at Palomar, by Oke. A brief report of his work will be given later. Ten-tative surface gravities are lower than expected, nearer $\log g = 7$ than 8 (expected for He interiors, $0.7 \, m_{\odot}$).

A recent study of the detectability of magnetic fields in DA stars by Preston (1970) involves quadratic Zeeman effect on the Balmer lines. Lines are both shifted to shorter wavelengths and broadened,

$$\langle \Delta \lambda \rangle = - 7.5 \times 10^{-23} \lambda^2 n^4 H^2, \tag{3}$$

where λ is in Å, and H in G. The run with principal quantum number, n, is such that at 10^6 G, the apparent velocity ranges from -11 km sec^{-1} at Hα, to -61 at Hγ, to -213 at Hε. These negative means are contradicted by observation, as is the trend with n. Note that if magnetic flux is conserved while a star contracts, H is proportional to R^{-2}, so the shift varies as the luminosity L^{-2}. In a typical DA with $L = 10^{-3} \, L_{\odot}$, the field is increased by 10^6 i.e., the initial field should be less than 1 G, or observable quadratic Zeeman effects would occur.

Kemp (1970a, b) has developed a theory of blackbody emission in a strong magnetic field, and predicts a circular polarization

$$q(\omega) \approx - eHm^{-1}\omega^{-1}. \tag{4}$$

This would be about 0.1% at 10^6 G. Laboratory experiments on metals showed the effect, insulators did not, and the carbon flame gave an opposite sign, ascribed to C_2 (Kemp et al., 1970a). Kemp found the circular polarization in EG 129 which is, appro-ximately 1 to 4% in the visible, and larger in the infrared. The process remains obscure; The Zeeman components of any frequency are unequal in strength in an optically thin gas. Angel and Landstreet (1970) found no linear polarization in several DA stars. Kemp claims a projected field strength of 10^7 G is present; extensive observations of EG 129 are underway.

Two white dwarfs vary quasi-periodically in brightness in roughly 10^3 sec; HL Tau 76, EG 165 was claimed to have a variable spectrum, but from several spectra seems to be a simple DA; G44-32, EG 72, is 17th mag, but one spectrum seems to be that of a DC. Recent spectra of some unusual white dwarfs are shown in Figure 3, from microphotometer tracings on an intensity scale. They have to be normalized for the sensitivity of the system, but are sufficient to locate features of these unusual types of stars.

D. RED SUBLUMINOUS STARS

In an extensive series of papers Eggen (1968, 1969, 1970) has shown that G-K-M stars of high ultraviolet excess, or suspected very high space motion, may include genuinely subluminous stars. Some parallaxes are available and some have common proper motion main-sequence stars. The red 'subdwarfs' are an extension of the well-known sdF and SdG stars of very low-metal content, but (1) seem to have excessively large $\delta(U-B)$ and (2) when calibrated by $R-I$ (in which line-blocking should be relatively small) they fall below the Hyades M_I, $R-I$ sequence. At first it seemed plausible that these were red degenerate stars. By 1968, I already found that the yield of classical red white dwarfs was very low, and that the number of newly discovered

TABLE II
Southern LTT stars, f, g, k, Eggen

(A) Greenstein, Published Kumar's *Low-Luminosity Stars*, 20 Stars
Types sdG-sdM; $\langle T \rangle = 520$ km sec^{-1}; $\langle \delta(U-B) \rangle = 0.31$ mag;
Two degenerate stars found, both bluer than the sun

LTT	B − V	U − B	$\delta(U-B)$	T	Type	Remarks
375	0.65	− 0.32	0.50	1800	DC	EG 246 (new)
7983	0.23	− 0.61	0.71	1100	DA	EG 137 (old)

(B) Other LTT stars, large T or $\delta(U-B)$

LTT	B − V	U − B	$\delta(U-B)$	T	Type	Remarks
2066	0.57	− 0.21	0.31	500	sdG	Extr. wk. CH
2981	0.66	− 0.18	0.38	3300	DC	L97-12 = EG 56 (old)
5454	0.68	− 0.16	0.38	350	sdG	Extr. wk. CH stg., vel. = + 223
5560	0.73	− 0.09	0.39	350:	sdG	Extr. wk. CH stg., vel. = − 216
5852	0.62	− 0.20	0.36	280	sdG	Extr. wk., vel. = − 35
6079	0.56	− 0.23	0.32	WD, 690	sdG	Extr. wk.
6194	0.52	− 0.13	0.16	710	sdF	Extr. wk.
6307	0.84	+ 0.10	0.43	200:	sdG	Extr. wk.
7132	0.72	− 0.13	0.41	230	sdG	Extr. wk. CH stg. LPM 661, vel. = − 216
7238	1.04	+ 0.53	0.39	200:	idK	CH stg.
7381	1.36	+ 1.25	0.00	700:	sdM	MgH stg. Metals wk.
12560	0.41	− 0.23	0.24	400	sdF	Ross 889, vel. = − 59
13746	0.64	− 0.09	0.27	4500	DC	EG 95 (old)
14272	0.86	+ 0.27	0.27	300:	sdK	

(C) Lowell stars, large T or $\delta(U-B)$

LTT	B − V	U − B	$\delta(U-B)$	T	Type	Remarks
G39-27	0.65	− 0.06	0.06	1200	DC	EG 40 (old)
G99-44	1.06	+ 0.81	0.18	2800	DK	EG 45 (old)
G113-40	0.93	+ 0.51	0.19	500	sdK	CH, Mg I stg. LTT 3144
G128-7	0.67	− 0.16	0.37	3600	DAss	GR 284 (new)
G134-22	0.73	+ 0.02	0.28	3600	DC	EG 16 (old)

$\langle T \rangle = 420$ km sec^{-1}; $\langle \delta(U-B) \rangle = 0.30$; $\langle m \rangle = 13.0$

Totals, old and new: non-degenerate 33 stars; degenerate 8; new degenerate 2.

stars like Wolf 489 was zero (Greenstein, 1969b). Table II, III, IV herewith show the type of evidence, and the spectroscopic results.

What do we expect to see? First, if stars have $\log g = +7.4$, (as do low mass DA stars) i.e., $R = 0.03\,R_\odot$, and have a temperature near that of the sun $M_v \approx +12$. They have very low opacity atmospheres; and unless the metal abundance is high, Rayleigh scattering will dominate. Then we expect an ultraviolet deficiency, not an excess. Next, if there are some metals, and H^- dominates, various degrees of line-blocking by broad invisible lines might occur, probably resulting in an approach to the black-body line. If $\log g$ is even lower and the atmosphere has a low pressure region containing metals, individual lines might be seen.

Opacity and line broadening vary with composition, electron and gas pressure; cool stars would be expected, near the main sequence, to show effects of both the metal/hydrogen ratio and g (roughly as the square root). Although a variety of simple formulas have been derived which depend on the fractional ionization of the metals, and the opacity dependence on electron or gas pressure, these would not be profitable here. Over a modest range of g, we expect strong lines to show larger pressure broadening; if the metals are very deficient and the opacity low, this would be easily detected. In case Rayleigh scattering or molecules contribute to the opacity, g would hardly affect the lines. We do not even know whether Eggen's red subluminous stars have high g, since it is conceivable that they have low mass.

TABLE III

Eggen's suspected red degenerate stars;
four Lowell fields, color class $+ 1$

Star	B − V	$\delta(U - B)$	T (km sec⁻¹)	Type	Remarks
G138-4	0.48	0.19	530	sdG	CH stg. H 14
-6	0.40	0.26	790	sdG	Extr. wk. H 13, vel. $= -258$
-16	0.83	0.23	920:	sdG	CH stg.
-21	1.11	0.16	440:	sdMp	TiO wk.
-38	0.97	0.06	630:	idK	CH stg.
-53	0.93	0.30	540:	idK	
-65	0.87	− 0.13	800:	idK	CH stg. vel. $= +70$
G141-6	0.87	0.18	1000	sdK	Extr. wk.
-15	0.52	0.20	470	sdG	Extr. wk. H 15, CH wk., vel. $= -290$
G181-19	0.81	0.34	WD, 310:	sdK	Extr. wk. CH stg., vel. $= -172$
-45	0.74	0.36	WD, 180:	sdK	
G178-9	0.83	0.41	WD, 230:	sdK	Vel. $= -170$
-30	0.86	0.27	WD, 130	sdK	
-41	0.48	0.24	470	sdG	Extr. wk. CH wk. H 13
-49	0.98	0.31	WD, 200:	idK	Vel. $= -194$

$\langle m \rangle = 13.5$
15 Stars: none degenerate
$\langle T \rangle = 510$ km sec⁻¹ $\langle \delta(U - B) \rangle = 0.23$

Total, Tables II, III: 48 non-degenerate, 8 degenerate; 6 are rediscoveries; 2 new degenerate stars.
$\langle T \rangle = 450$ km sec⁻¹ $\langle \delta(U - B) \rangle = 0.28$

Degenerates: $\langle T \rangle = 2750$ km sec⁻¹; $\langle \delta(U - B) \rangle = 0.34$

TABLE IV

New possible red degenerate stars

$T > 200$ km sec^{-1}, Large $\delta(U - B)$, or $\delta(U - B) > 0.20$ at $(B - V) > 0.70$

Star	B − V	$\delta(U - B)$	T	Type	Remarks
G71-53	0.77	0.36	290	idK	Eggen 530; CH stg.
VA 216	0.96	0.03	340	dM	Suggested DM in Hyades; TiO, metals seen.
VA 391	0.70	0.17	290	sdF	Suggested DK in Hyades; Ca II sharp.
G47-48	0.57	0.02	900	dK	
GD 103	0.75	0.29	190	sdK	
GD 105	0.70	0.23	220	sdK	Extr. wk., CH stg.
GD 118	0.45	0.25	360	sdG	
G119-11	1.11	0.31	900:	sdK	Could be DK.
G119-57	1.46	–	410:	idM	
G61-35	1.12	0.20	510:	sdK	
G164-61	0.59	0.19	540	sdG	Extr. wk., CH stg.
LTT 6660	0.87	0.45	160:	sdK	Extr. wk.
G17-37	0.84	0.40	240:	sdK	Extr. wk., CH stg.
GD 211	0.64	0.24	330	sdK	Extr. wk., CH stg.
G155-27	1.36	–	740:	sdM	Eggen 700; wk. lines; MgH stg.
GD 214	0.81	0.38	140:	sdG	Vel. = − 67
GD 217	0.70	0.39	80:	sdG	Vel. = + 96
GD 224	0.88	0.23	160	sdK	
G125-59	1.03	0.19	400:	sdK	Extr. wk.; Vel. = − 277.
G18-51	1.42	–	550:	sdM	CC 1363; MgH present; Vel. = − 157.

Fainter Sample $\langle m_V \rangle = 14.5$

20 Stars $\langle T \rangle = 390$ km sec^{-1}; $\langle \delta(U - B) \rangle = 0.25$

17 Non-degenerate, 1 Possible, 2 improbable degenerates.

The clues searched for in the spectra so far obtained have been: (a) presence of extreme line broadening, or enhancement of strong lines with respect to weak ones; (b) presence of abnormal hydrogen lines; (c) presence of abnormal molecular bands; (d) discrepancies between the ionization level and the color; (e) abnormally low excitation level. In general, the tables indicate that the red subluminous and subdwarf stars have: (a) abnormally strong, sharp hydrogen lines for their color; (b) abnormally weak neutral and ionized metallic lines; (c) weak TiO, and very strong MgH bands; (d) very strong CH bands in weak-metal stars.

The statistical results are that the percentage of classical red degenerate stars is nearly zero. The statistics are:

(A) Table II shows 33 non-degenerate; 2 new degenerate of which one is like vMa2, but shows no lines, and one is DA; and rediscovery of 6 EG yellow degenerate stars.

(B) Table III shows 15 non-degenerate stars.

(C) Table IV shows 20 non-degenerate stars.

There are 68 non-degenerate stars, 1 DA, 1 yellow DC found, and 6 old objects rediscovered. No analogs of the intrinsically faint yellow and red stars like vMa2, W 457, W 489 were found. Genuine 'red-white dwarfs', with 0.01 $R_\odot$ are rare in space and make up less than 1% of the proper-motion stars. But these proper-motion

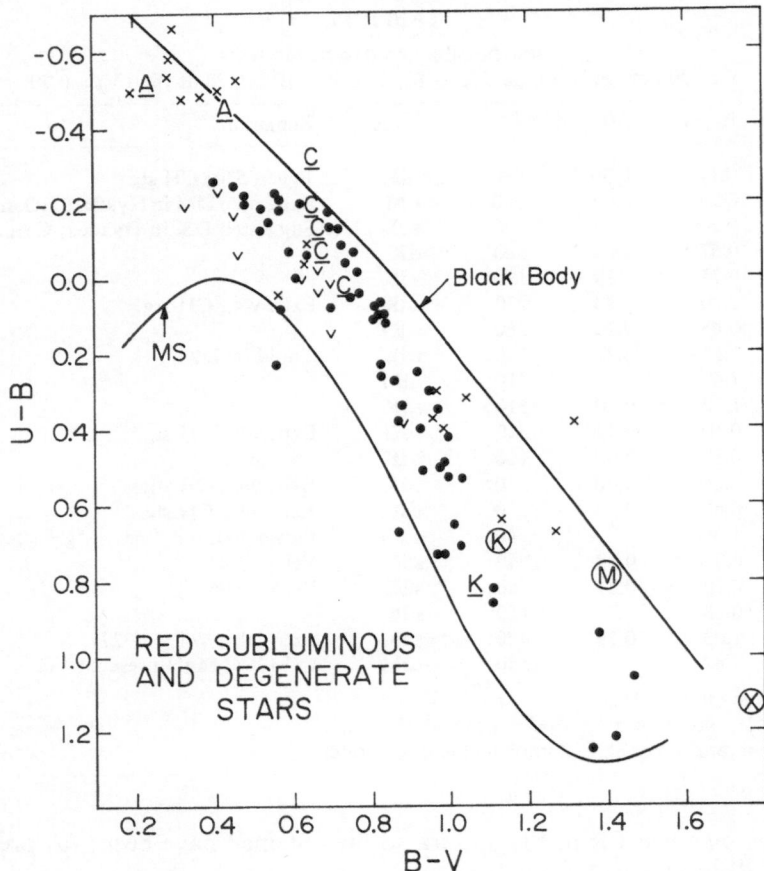

Fig. 4. Results of search for red degenerates among objects of large $\delta(U-B)$ and transverse motion. Failures (checks) are stars with $\delta(U-B) < 0.25$, $T < 300$ km sec^{-1}. Dots are extreme subdwarfs (F, G, K, M) above these limits. Degenerate stars are shown by crosses if previously known, underlined, if found in this search, and circled if possibly subdwarf or degenerate. C̲ are DC Stars.

stars, as a group, have luminosities (from UBV) such that the tangential velocity is 450 km sec^{-1} (Tables II, III) or 390 km sec^{-1} (Table IV). Further, the excesses for sdG-sdM stars average $\langle\delta(U-B)\rangle = 0.28$ mag (Tables II, III), and 0.25 (Table IV). The degenerate stars in this sample had $\langle T \rangle = 2750$ km sec^{-1} and $\langle\delta(U-B)\rangle = 0.34$. Thus only stars with very extreme $T \geqslant 1000$ km sec^{-1} derived from UBV photometric parallaxes are likely to be classical yellow or red degenerates.

Figure 4 illustrates these results based largely on Eggen's (1968, 1969) photometry. I have not been able to observe his more recently published stars. The sample is quite large enough to show how few new red degenerate stars are found, but it does concern a relatively bright group of stars (about 14th mag), which are very peculiar. The fainter group (Table IV) is no less peculiar. The large values of $\delta(U-B)$ in cool stars were not originally expected, but persist to sdM type. The very large tangential velocity (over 400 km sec^{-1}) is that of extreme halo population II. The radial velocities obtained at 90 or 190 Å mm^{-1} are of low accuracy, but as tables II–IV show, are very large; for 23 stars $\langle|\varrho|\rangle = 142$ km sec^{-1}, while $\langle\varrho\rangle = -93$ km sec^{-1}. The large negative mean radial component arises from the distribution of the stars over the sky with respect to the direction of galactic rotation. Among 14 stars in Tables II–IV, 11 have large negative radial velocities, and 3 positive. But evaluating the contribution to the V-velocity, in the direction of galactic rotation, we find 9 stars have a radial velocity contribution, $\varrho(\partial V/\partial\varrho)$, to V of -120, and 5 have $\varrho(\partial V/\partial\varrho) = +30$. Thus, in these stars, the high negative radial velocity is required to keep the stars bound in the galaxy. The radial velocity contribution to the W-coordinate, $\varrho(\partial W/\partial\varrho)$ averages 90 km sec^{-1}, so that even with zero proper motion the stars would be halo objects.

The tangential motion for these stars is 400 km sec^{-1}, very large compared to the radial-velocity dispersion; clearly it is possible to reduce it to a reasonable value if the stars are closer to us i.e., have lower luminosity than given by the UBV photometric parallax. Thus, if $\langle|\varrho|\rangle$ is equated to $2/\pi\langle T\rangle$, which would be true for stars uniformly distributed over the sky, and with random space motions, we find that $\langle T\rangle$ should be about 220 km sec^{-1} or about one-half what we observe. Then the stars are 1.5 mag below the Hyades main sequence i.e., half the main sequence radius, or one-quarter the surface gravity. (This was already indicated by Greenstein (1969b).) Quantitative spectroscopic analysis, using H lines, Ca I and Fe I line-broadening theory and models should be sufficient to reveal so large a factor. Thus Eggen's stars seem to be subluminous, but not normal red degenerate stars. They are far above the low-mass DA white dwarfs which might have 0.03 $R_\odot$. The existence of low-mass hydrogen-rich degenerate stars below 0.1 $M_\odot$ has not been established observationally; their radii would be given from non-relativistic partially degenerate models, and might well be greater than 0.10 $R_\odot$. It seems improbable that they would be as large as 0.50 $R_\odot$, and very improbable that they would be as hot as these stars.

The fact that genuine red degenerates have not been found in this study, while several are known close to the sun depends on the high apparent brightness ($\langle V\rangle \approx 14.5$) of the sample. However, it is clear that the red degenerates are, in fact, not very common, and certainly not in proportion to their cooling time based on a simple theory. Solidification and low specific heat are responsible (see Greenstein 1969c) for shortening cooling times at low T_{eff}. Where can red degenerates be found? Luyten (1970) has published an extensive list of 1055 faint stars of large proper motion: he finds that about 70% are of color class k or redder. The reduced proper motion H is related to the tangential motion by

$$H = m + 5 + 5\log\mu = M + 5\log T - 5\log 4.74. \tag{5}$$

His scale of apparent magnitudes is photographic, so we must subtract 1.8 mag to obtain the visual luminosity

$$M_v \approx H_{pg} + 1.6 - 5\log T. \tag{6}$$

Stars near $m_{pg} = 20$ exist with $\mu = 1''$ yr^{-1}, and many m stars have $H_{pg} \gtrsim 22$. If $T = 75$ km sec^{-1}, as Luyten adopts, (valid for old disk stars and moderately high-velocity stars) $M_v \gtrsim +14$ i.e., these faint objects can be either main sequence or red degenerates.

For $T = 400$ km sec^{-1}, however, as in the sample discussed in this paper, $M_v \gtrsim +11$, the stars can be M dwarfs, but *not* red degenerates. For statistical purposes, radial-velocity dispersions of red stars of large H might provide a clue as to the fraction of degenerate stars. Color discriminants might be found from a combination of B, V, R, I and further infrared photometry. Differences between known red degenerates and main-sequence stars, if recognizable, would enormously simplify spectroscopic search for the faint end of both groups of objects.

Acknowledgements

This research was supported in part by the U.S. Air Force under grant AFOSR 68-1401 monitored by the Air Force Office of Scientific Research of the Office of Aerospace Research.

References

Angel, J. R. P. and Landstreet, J. D.: 1970, *Astrophys. J. (Letters)* **160**, L147.
Eggen, O. J.: 1968, *Astrophys. J. Suppl.* **16**, 97.
Eggen, O. J.: 1969, *Astrophys. J. Suppl.* **19**, 31.
Eggen, O. J.: 1970, in press.
Eggen, O. J. and Greenstein, J. L.: 1965a, *Astrophys. J.* **141**, 93. (EG I)
Eggen, O. J. and Greenstein, J. L.: 1965b, *Astrophys. J.* **142**, 925. (EG II)
Eggen, O. J. and Greenstein, J. L.: 1967, *Astrophys. J.* **150**, 927. (EG III)
Greenstein, J. L.: 1966, *Astrophys. J.* **144**, 496.
Greenstein, J. L.: 1969a, *Astrophys. J.* **158**, 281. (EG IV)
Greenstein, J. L.: 1969b, *Low-Luminosity Stars* (ed. by S. S. Kumar), Gordon and Breach, New York, p. 281.
Greenstein, J. L.: 1969c, *Comments Astrophys. Space Phys.* **1**, 62.
Greenstein, J. L.: 1970, *Astrophys. J. (Letters)* **162**, L55. (EG VI)
Greenstein, J. L. and Matthews, M.: 1957, *Astrophys. J.* **126**, 14.
Greenstein, J. L. and Eggen, O. J.: 1966, *Vistas in Astronomy* (ed. by A. Beer), Pergamon Press, London, p. 63.
Kemp, J. C.: 1970a, *Astrophys. J.* **162**, 169.
Kemp, J. C.: 1970b, *Astrophys. J. (Letters)* **162**, L69.
Kemp, J. C., Swedlund, J. B., and Evans, B. D.: 1970a, *Phys. Rev. Letters* **24**, 1211.
Kemp, J. C., Swedlund, J. B., Landstreet, J. D., and Angel, J. R. P.: 1970b, *Astrophys. J. (Letters)* **161**, L77.
Luyten, W. J.: 1970, *The Stars of Low Luminosity*, University of Minnesota Press, Minneapolis.
Mihalas, D.: 1965, *Astrophys. J. Suppl.* **9**, 321.
Newell, E. B.: 1969, Thesis, Australian National University.
Ostriker, J. P. and Bodenheimer, P.: 1968, *Astrophys. J.* **151**, 1089.
Preston, G. W.: 1970, *Astrophys. J. (Letters)* **160**, L143.
Sargent, W. L. W. and Searle, L.: 1968, *Astrophys. J.* **152**, 443.
Searle, L. and Rodgers, A. W.: 1966, *Astrophys. J.* **143**, 809.
Stephenson, C. B., Sanduleak, N., and Hoffleit, D.: 1968, *Publ. Astron. Soc. Pacific* **80**, 92.
Stoeckly, R. and Greenstein, J. L.: 1968, *Astrophys. J.* **154**, 909.
Strom, S.: 1970, private communication.
Terashita, Y. and Matsushima, S.: 1969, *Astrophys. J.* **156**, 203.
Wickramasinghe, D. T. and Strittmatter, P. A.: 1970, *Monthly Notices Roy. Astron. Soc.* **147**, 123.

10. OBJECTIVE PRISM SURVEYS FOR NEW WHITE DWARFS

C. B. STEPHENSON

Warner and Swasey Observatory,
Case Western Reserve University, East Cleveland, O., U.S.A.

Abstract. The spectroscopic criteria for identifying strong-lined DA-type stars on low-dispersion objective prism plates are reviewed, as are the objective prism surveys made to date with discovery of white dwarfs a partial or major objective (Table I). It is claimed that there may be upwards of 100 undiscovered white dwarfs brighter than B magnitude 14.0. Some data for all the white dwarfs so far discovered by objective prism are collected in Table II; new identification charts are given for two of these.

The only type of white dwarf capable of secure spectroscopic recognition on objective prism plates of low dispersion (say ~ 600 Å/mm) is the class of DA stars – fortunately, a rather populous class – having strong hydrogen lines. As Figure 1 shows, at such spectral dispersions these stars somewhat resemble normal A-type dwarfs, with the following exceptions: the Balmer lines are deep and very broad, and disappear with Hε or Hζ; there is no perceptible Balmer discontinuity; and there are no other lines.

The spectral properties just mentioned are fairly well understood theoretically, and for us the only remaining spectroscopic points of interest are the known sources of possible spectral classification error. These are as follows. First, the weak-lined DA stars may easily be confused with rapidly rotating B stars, unless the rapid rotator betrays itself either by the presence of line or Balmer continuum emission or by a significant Balmer discontinuity. But since the Bnn stars can never have deep lines, they can of course never mimic the *strong-lined* DA stars. Second, and last, a close overlap of the objective prism spectra of two early A-type or late B-type dwarfs, with slight separation in the direction of spectral dispersion, resembles a DA spectrum in the blue region. However, this is a possible cause of false DA classification only if the plate lacks the ultraviolet.

Granted that safe identification of strong-lined DA stars is possible in the manner outlined, the fact that these plates reach only to magnitude 13 or 14 would seem to make white dwarf discovery by objective prism rather limited in power in comparison, say, with proper motion surveys. Nevertheless about half of the 40-odd DA stars brighter than B magnitude 14.0 were discovered from their colors or objective prism spectra, notwithstanding that the sky has been more thoroughly surveyed for large proper motions to this magnitude limit than for colors or spectral types. It appears from the data of Eggen and Greenstein (1965) that the white dwarfs found in proper motion surveys tend to be high-velocity stars; the mean space motion for such DA stars is nearly 80 km/sec and even for those brighter than B magnitude 14.0 it is nearly 50 km/sec. Nevertheless there are large numbers of low-velocity white dwarfs (cf. Greenstein, 1965a). To magnitude 14 there may well remain more undiscovered white dwarfs than those so far found in proper motion surveys. Indeed the third brightest white dwarf then known was discovered as late as 1967 (Stephenson *et al.*,

Luyten (ed.), White Dwarfs, 61–66. All Rights Reserved.
Copyright © 1971 *by the IAU.*

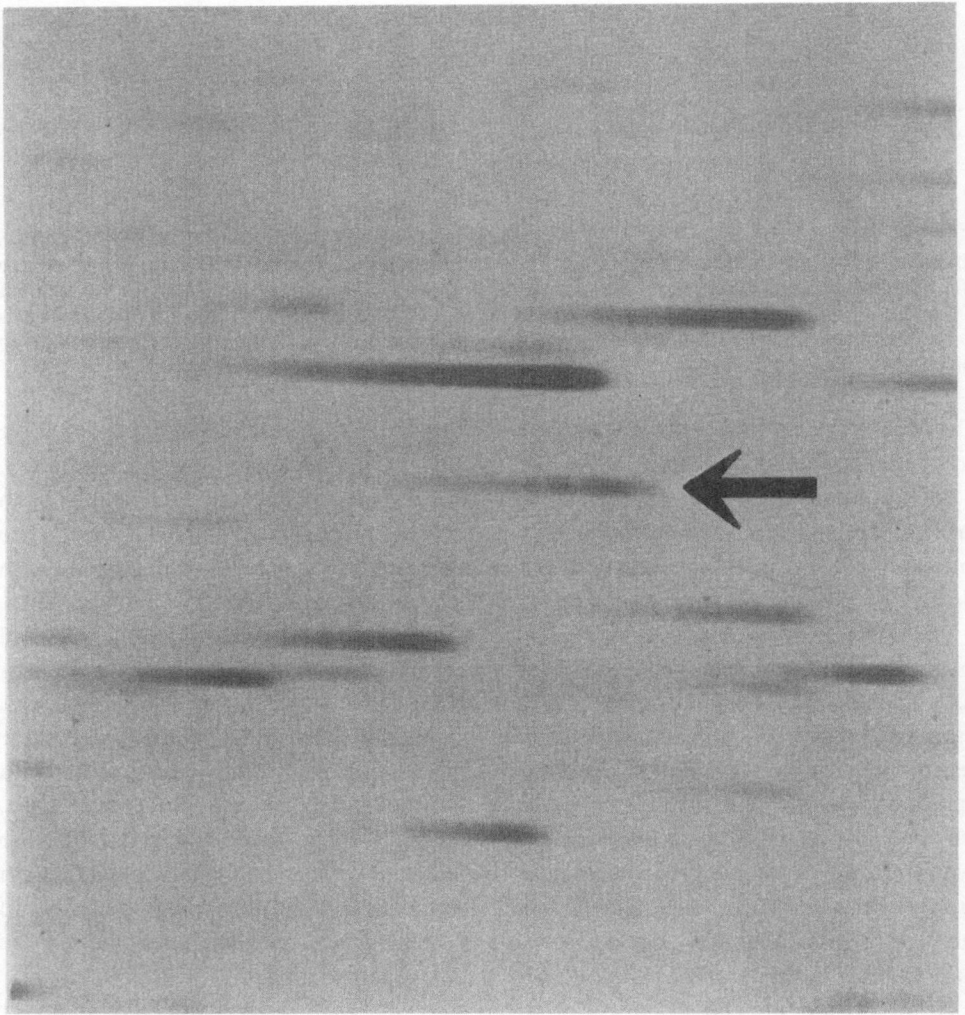

Fig. 1.   Portion of a Warner and Swasey Observatory objective prism plate of the field of the DA-type white dwarf Greenwich Astrographic +73°8031 (arrow). Original dispersion 580 Å/mm at Hγ; red limit ∼ 4900 Å, on the right, and blue limit ∼ 3300 Å. North is to the left and east below. For comparison, note the B star northeast of the white dwarf. The B magnitude of the white dwarf is 12.9 (Eggen and Greenstein, 1965).

1968) from its objective prism spectrum and another, even brighter though slightly less certain, was similarly discovered a little later (Bidelman, 1968). One may also conclude that there may be a substantial number, say 50 or so, undiscovered bright white dwarfs on the basis of Luyten's discussion of the space density of white dwarfs (Luyten, 1958). This conclusion does not by any means lessen the importance of the vital proper motion surveys of the past and present, but it does make the subject of this paper more than academic.

TABLE I

Objective prism surveys to date with discovery of DA stars as partial or major objective

General area covered	Approx. area in sq. deg.	Limiting B mag.	No. of DA stars found	Approx. B mag.* of DA's found	DA paper	Notes
Luminous Stars in the Northern Milky Way IV and VI [1, 2]	2500	12.0–12.5	0			
Luminous Stars in the Southern Milky Way [3]	6000	11.5–13.0	1	10.8	[4]	
Ursa Major	270	13.5–14.0	3	13.5, 13.8, 13.9	[5]	a
Cleveland zenith	800	13 –14	2	13.4, 13.6	[6]	
Michigan survey of southern hemisphere	14000	10	1	9.8:	[7]	b
South galactic polar cap	230	13.5	2	12.9, 13.1	[8]	c

* Discoverers' estimates, like the limiting magnitudes of the surveys.

Notes to Table I: (a) One of these stars is LB 253, previously known as a blue star but not as a white dwarf. (b) If a true white dwarf, this star is of the weak-hydrogen type. This survey still in progress. (c) One of these stars is Tonantzintla 191 = BPM 46931, previously known as a blue star of measurable proper motion but not as a white dwarf.

References cited in Cols. 1 and 6: [1] Nassau and Stephenson, 1963; [2] Nassau et al., 1965; [3] Stephenson and Sanduleak, in preparation; [4] Stephenson et al., 1968; [5] Stephenson, 1960; [6] Stephenson, 1962; [7] Bidelman, 1968; [8] Sanduleak and Philip, 1968.

TABLE II

Data for individual white dwarfs discovered by objective prism

Name	α (1900)	δ	V	B−V	U−B	Source of photometry	DA paper	Total proper motion (source)	Published slit observation	Notes
T 191 = BPM 46931	0^{h}50^m.9	−33°16′	13.36	−0.25	−1.07	[1]	[1]	0″.15 (2)		a
	1 02.4	−33 54	13.56	−0.28	−1.10	[1]	[1]	0″.24 (5)		b
LB 253	11 01.8	+60 31	13.80	−0.02	−0.81	[3]	[4]		[3]	c
C1	12 10.8	+53 04	13.34	+0.53	−0.48	[3]	[4]	0″.11 (5)		c, d
AC + 58°43662	13 42.4	+57 30		+0.55	−0.27	[3]	[4]	0″.29 (4)		e
C2	16 05.0	+42 21	13.85	+0.06	−0.54	[3]	[6]		[3]	f
CD−38°10980	16 16.8	−39 00	10.97	−0.14	−0.96	[7]	[8]	0″.08 (8)	[7]	g
CD−42°14462	19 40.7	−42 15	9.7			CD	[9]	0″.04 (10)		h
C3	23 26.8	+40 28	13.82	+0.03	−0.72	[3]	[6]		[3]	i

Notes to Table II: (a) Chart in ref. [1]. (b) Chart by Chavira, 1958. (c) Chart in Figure 2. (d) Has emission lines, presumed due to a dMe companion, according to Greenstein, 1965b. A recent infrared (Kodak I − N emulsion behind Schott RG8 filter) objective prism plate shows the M star to be of spectral type about M 2. If the dM star has a visual absolute magnitude $M_v = 10.1$ and V − I = +2.4, then my approximate calibration of these plates would give for the DA star $M_v = 9.8 \pm 1$:. (e) The published V magnitude has to be a misprint; perhaps should be 13.73. The right ascension published in ref. [4], and repeated in ref. [3], is also a misprint; see the erratum cited with Stephenson, 1960. (f) Chart in ref. [6]. (g) Chart in ref. [8]. This star has a G-type dwarf companion, see Alexander and Lourens, 1969. As they point out, the G star has a published trigonometric parallax of 0″.049 ±0″.010, which would give the DA star a visual absolute magnitude of 9.5. (h) Weak lines; region of Balmer limit unobserved, but according to Bidelman probably not Bnn because has no Hα emission. (i) Chart in ref. [6].

References cited in Cols. 7–10: [1] Sanduleak and Philip, 1968; [2] Luyten, 1961; [3] Eggen and Greenstein, 1965; [4] Stephenson, 1960; [5] Luyten (1969) A Search for Faint Blue Stars, vol. L; [6] Stephenson, 1962; [7] Hiltner et al., 1968; [8] Stephenson et al., 1968; [9] Bond, 1968; [10] Russo, 1969.

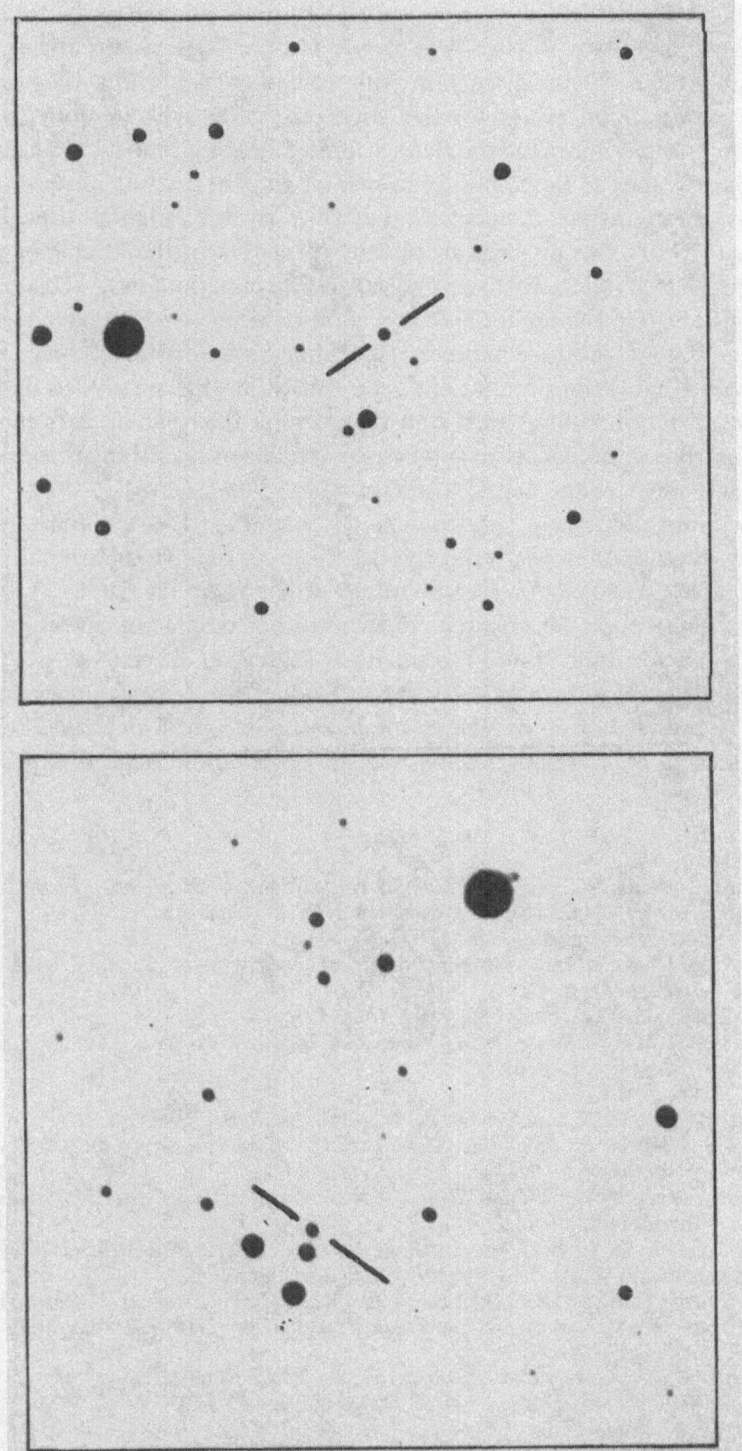

Fig. 2. Identification charts for the two white dwarfs of Table II that lack previously published ones, reproduced from the *Lick Sky Atlas*. North is at top and east to the left, and the fields are about 25' ×25'. Left: LB 253. The brightest star in the field, in the southwest quadrant, is BD + 60°1309. Right: C1. The brightest star in the field, north of center, is BD + 53°1535.

Table I summarizes the objective prism surveys to date, known to me, that have been made with the discovery of new white dwarfs as a partial or major objective. All were done at about 600 Å/mm dispersion with equipment recording the spectrum well below the Balmer limit, except for the Michigan survey which employed much higher dispersion and did not reach the Balmer limit. While the statistical uncertainty is immense, at face value the Luminous Stars surveys suggest that the unsurveyed sky might contain something like 5 undiscovered white dwarfs brighter than twelfth magnitude, one of which has presumably already turned up in the Michigan survey. The Ursa Major-Cleveland zenith surveys suggest over a hundred brighter than fourteenth magnitude. On the other hand, any additional surveys that have remained unknown to me would lower these numbers, since they have not yielded further white dwarfs. The present indications are, at any rate, that either the surveys to date have been very lucky or that our white dwarf statistics even for the brighter stars can stand improvement and that objective prism surveys are capable of furnishing this improvement in the case of the strong-lined DA stars.

Another possibility, that these spectroscopically identified DA stars are not true white dwarfs, is in my opinion very unlikely, but this does make it additionally desirable for trigonometric parallax observers to observe the stars so far found. The photometric and slit-spectroscopic observations of them to date confirm their white dwarf nature (Eggen and Greenstein, 1965; Eggen, 1965; Hiltner et al., 1968).

Table II list additional information for the individual white dwarfs discovered to date by objective prism. For convenience, some of the data of Table I are repeated here. Stars possessing entries under 'published slit observation' have been observed with slit spectrograms.

References

Alexander, J. B. and Lourens, J. v. B.: 1969, *Monthly Notices Astron. Soc. South Africa* **28**, 95.

Bond, H. E.: 1968, *IAU Circ.*, No. 2120, communicated by W. P. Bidelman.

Chavira, E.: 1958, *Bol. Obs. Tonantzintla Tacubaya* **2**, No. 17, 15.

Eggen, O. J.: 1965, in *First Conference on Faint Blue Stars* (ed. by W. J. Luyten), University of Minnesota Press, Minneapolis, p. 47.

Eggen, O. J. and Greenstein, J. L.: 1965, *Astrophys. J.* **141**, 83.

Greenstein, J. L.: 1965a, in *First Conference on Faint Blue Stars* (ed. by W. L. Luyten), University of Minnesota Press, Minneapolis, p. 62.

Greenstein, J. L.: 1965b, *Loc. Cit.*, p. 97.

Hiltner, W. A., Stephenson, C. B., and Sanduleak, N.: 1968, *Astrophys. Letters* **2**, 153.

Luyten, W. J.: 1958, *A Search for Faint Blue Stars*, vol. X: *On the Frequency of White Dwarfs in Space*, Lund Press, Minneapolis.

Luyten, W. J.: 1961, *Bruce Proper Motion Survey: The General Catalogue*, part E, University of Minnesota Press, Minneapolis.

Nassau, J. J. and Stephenson, C. B.: 1963, *Luminous Stars in the Northern Milky Way*, vol. IV, Hamburger Sternwarte and Warner and Swasey Observatory, Hamburg.

Nassau, J. J., Stephenson, C. B., and MacConnell, D. J.: 1965, *Luminous Stars in the Northern Milky Way*, vol. VI, Hamburger Sternwarte and Warner and Swasey Observatory, Hamburg.

Russo, T. W.: 1969, *IAU Circ.*, No. 2128.

Sanduleak, N. and Philip, A. G. D.: 1968, *Publ. Astron. Soc. Pacific* **80**, 437.

Stephenson, C. B.: 1960, *Publ. Astron. Soc. Pacific* **72**, 387; erratum **73**, 247.

Stephenson, C. B.: 1962, *ibid.* **74**, 210.

Stephenson, C. B., Sanduleak, N., and Hoffleit, D.: 1968, *Publ. Astron. Soc. Pacific* **80**, 92.

11. EFFECTIVE TEMPERATURES OF WHITE DWARFS

J. B. OKE and H. L. SHIPMAN*

*Hale Observatories,
California Institute of Technology,
Carnegie Institution of Washington*

1. Introduction

White dwarf stars are among the most challenging and interesting objects which can be studied. Because they represent the interiors of highly-evolved stars, the chemical composition can be enormously variable from object to object. Furthermore, because of the very large gravities, the composition of the atmosphere may be very different from that in the interior. The theory of the degenerate interior provides a relation among mass, radius and chemical composition. Since temperatures, effective gravities, and redshifts can, for certain stars, provide further relations between mass and radius, one can hope to make checks on the theory which are not possible with ordinary stars.

Two parameters which are required, if the maximum possible information is to be obtained from white dwarfs, are effective temperature and gravity. These parameters are obtained for normal stars by matching absolute spectral energy distributions with fluxes computed from model atmospheres. In the case of white dwarfs, until a few years ago neither spectral energy distributions nor good model atmospheres existed. Effective temperatures had to be inferred from UBV photometry and interpolation between the main-sequence stars and black bodies. Further information was obtained from hydrogen-line profiles and rather simple model atmospheres (Weidemann, 1963).

2. Observations

It became evident two or three years ago that excellent model atmospheres would become available, and a program was launched to obtain spectral energy distributions of selected white dwarfs. The initial observations were made with the prime-focus scanner attached to the 200-inch Hale telescope. The practical magnitude limit was approximately 13, so only a few white dwarfs could be observed. With the completion of the 32-channel spectrometer two years ago (Oke, 1969) almost any known white dwarf could be observed. Consequently, the program was expanded to include several examples of the many different types of white dwarfs.

The spectral energy distributions cover the spectral range from 3200 Å to 10000 Å. Below 5600 Å the resolution is usually 80 Å; above 5600 Å is is 160 Å. The resolution is adequate to define the continuum and to measure equivalent widths of some features, but is not sufficient to measure profiles. A few scans have been made with higher

* NSF Graduate Fellow.

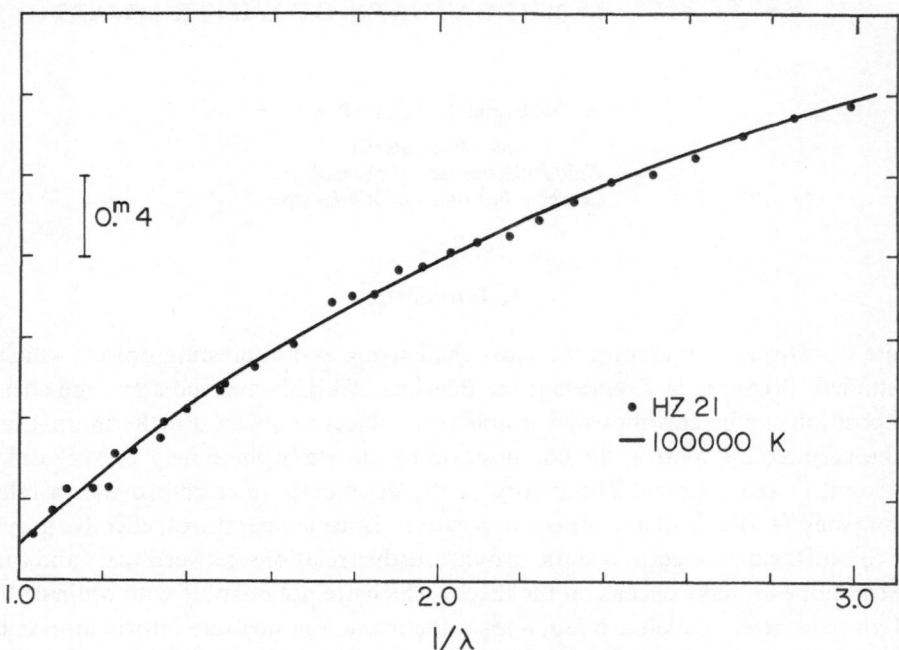

Fig. 1. *HZ 21*. This star has lines of H, HeI, HeII (Eggen and Greenstein, 1965). The continuum matches a black body with $T = 100000$ K. The absolute flux in magnitudes, as defined in Equation (3), is plotted against $1/\lambda(\mu)$.

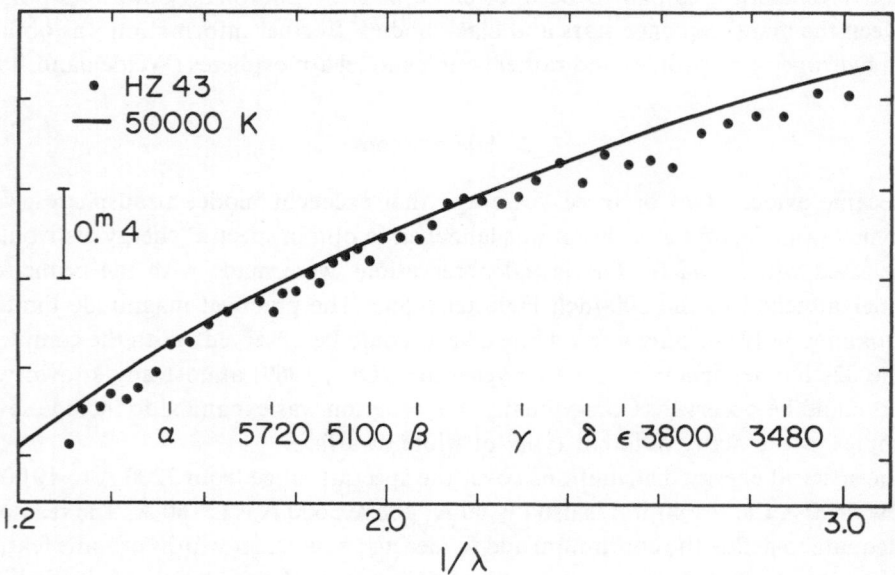

Fig. 2. *HZ 43*. This star has a faint red companion at a distance of 3″ which may contaminate the energy distribution slightly. The black body temperature is 50000 K. Hydrogen lines are weak (Eggen and Greenstein, 1965).

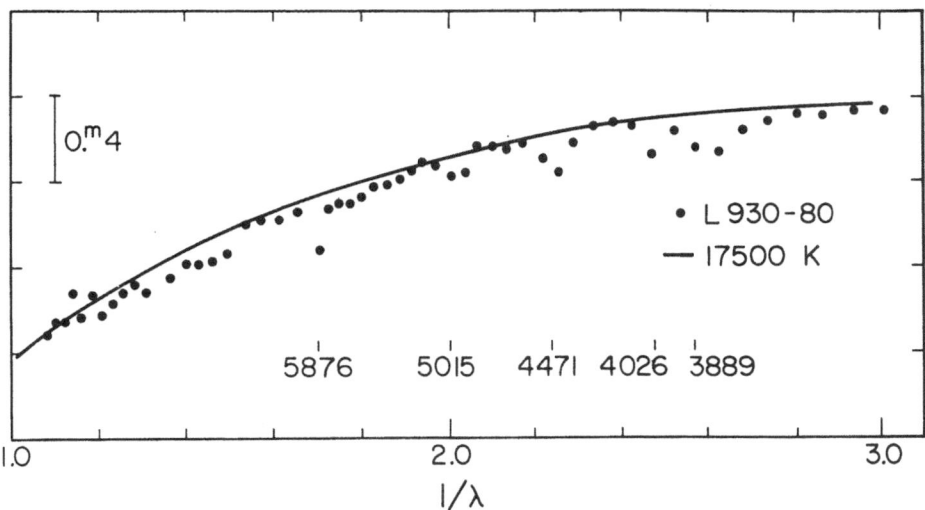

Fig. 3. *L 930-80* is of type DB and has strong Helium lines. The curve is a black body.

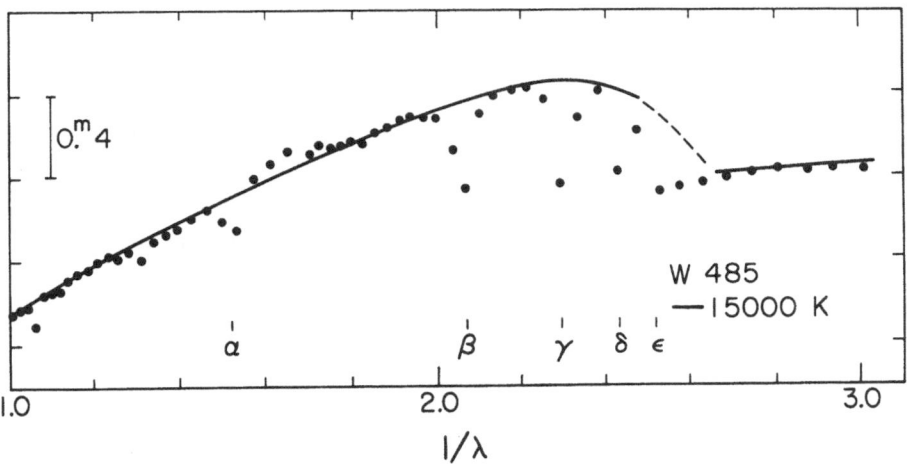

Fig. 4. *W 485* is a typical DA. The model is from Terashita and Matsushima (1969).

resolution and these can be used for profile measurements. Observations have been completed for approximately 35 white dwarfs. These include 13 of type DA, 7 of type DB, 4 of types DF and DG. The remainder are peculiar objects.

All of the objects studied here have been observed with a slit spectrograph by Greenstein (1958, 1960) and Eggen and Greenstein (1965, 1967), and these spectra can be used for line profiles and equivalent width measurements.

In Figures 1–5 are shown a few typical examples of observed energy distributions and the fitted model atmosphere fluxes (see below). In some cases, the fitted curves

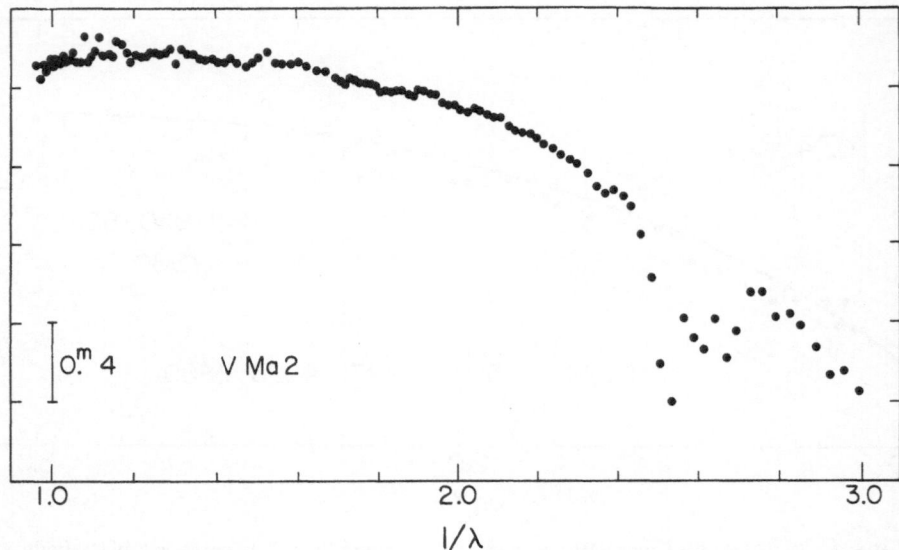

Fig. 5. *V Ma 2*. The best-known DG star. The effective temperature is probably less than 5000 K. The star has strong metal lines in the ultraviolet in addition to H and K of Ca II (Eggen and Greenstein, 1965).

are black-body curves. The results so far obtained from all the available stars are as follows:

(a) The DA stars have effective temperatures ranging from 13000 K to 50000 K.

(b) DB stars range from 15000 K to 25000 K.

(c) There are two stars with $T_e = 100000$ K.

(d) The DC star L 1363-3, which shows no features has a temperature in the neighborhood of 10000 K.

(e) DF and DG stars have a wide temperature range, at least from 4500 K to 9000 K.

3. White Dwarfs with Carbon Bands

It has been noted by Eggen and Greenstein (1967) that the white dwarf G 47-18 has both atomic C I lines and molecular bands of C_2. The energy distribution is shown in Figure 6 where the many C_2 bands are conspicuous. Another recently-discovered white dwarf, G 99-37, (Greenstein, private communication) clearly shows the same C_2 bands as G 47–18, but with different relative intensities. In addition, G 99-37 appears to have bands of CH. The difference in the band strengths and the presence of CH in one of the two stars may be a result of the obviously very different effective temperatures of the two stars.

4. Model Atmospheres

The model atmospheres used in this investigation were constructed using the program

ATLAS, written by R. Kurucz of the Harvard and the Smithsonian Observatories (cf. Strom and Avrett, 1964, 1965; Kurucz, 1969a, 1969b). Opacity sources included H I (continuous and line opacity), H^-, H_2^+, He I, He II, He^-, Si, Mg, C, N, O, Ne in appropriate stages of ionization, and electron and Rayleigh scattering. For these models, H was the only important opacity source. The chemical composition was taken

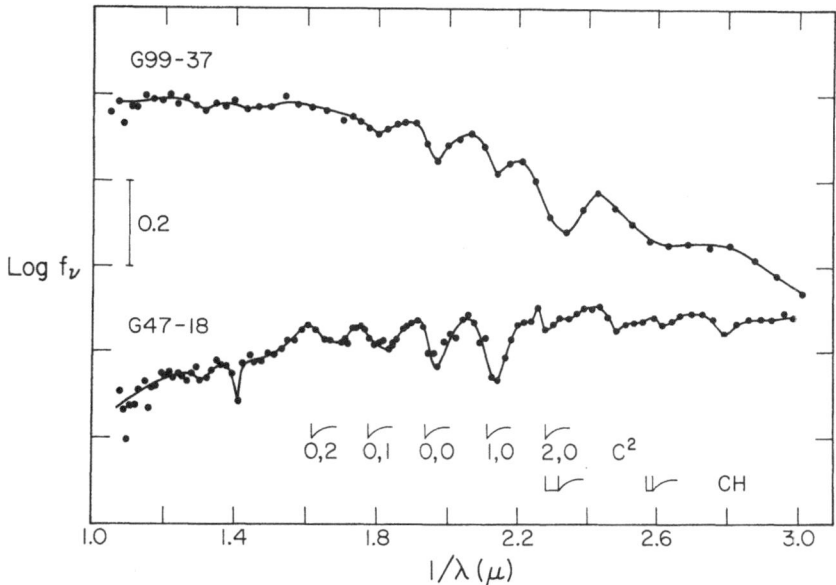

Fig. 6. Spectral energy distributions of G 47-18 and G 99-37. Various C_2 and CH molecular bands are indicated.

to be that of population I (He = 0.1, H = 0.9, metals = solar); although (see below) the chemical composition of DA atmospheres is actually almost pure H, the near-complete ionization of H and the dominance of H as an opacity source makes the metal abundance completely unimportant in its effect on the atmospheric structure. The effect of changing the He abundance to zero merely increases the surface gravity corresponding to a given model by 0.12 in the logarithm (cf. Terashita and Matsushima, 1969).

Model atmospheres were constructed for $T_{eff} = 12000$, 16000, 20000, and 25000 K and $\log g = 7$ and 8. One model with $T_{eff} = 16000$ K, $\log g = 7.5$ demonstrated that for the interpretation of the hydrogen-line strengths, linear interpolation over $\log g$ is satisfactory.

The present grid of models is quite similar to that presented by Terashita and Matsushima (1969). (Hereafter these authors will be referred to as TM).

The principal difference is that we have taken pressure ionization into account in computing the ionization equilibrium, white TM do not (cf. TM, 1966). This changes the electron densities at $\tau_{5000} \approx 1$ by a factor of about 2. The continuous energy distributions are unaffected by taking this factor into account, but the hydrogen-line strengths for a given model are *increased* over the earlier ones and consequently our surface gravities (and hence masses) are lower than those deduced by TM. The expression we used to compute the lowering of the ionization potential is

$$\Delta E = \frac{e^2}{\lambda_D}; \quad \lambda_D = \left[\frac{kT}{4\pi e^2 (n_e + n_i)}\right]^{1/2}.$$

Where λ_D is the Debye length; $(n_e + n_i)$ is the total charge density. At a temperature of 20000 K, this corresponds to

$$6 \log n = 21.64 - \log n_e$$

where n is the highest bound level of H. This is in reasonably good agreement with the results of Schatzman (1958) and Kolesov (1964), and agrees with the equation used by TM (1966) to compute the continuous energy distributions. A sample fit of the continuum is shown in Figure 4.

5. Procedure for Comparing Observations and Models

The scans described above for DA stars were fitted to the model atmospheres to determine the effective temperature. The deduced effective temperatures are plotted against the $(U-V)$ color (from Eggen and Greenstein, 1965) in Figure 7. It is apparent that for DA stars cooler than $T_{eff} = 25000$ K, $U-V$ is an unequivocal indicator of effective temperature. The difference between our temperature scale and that of Terashita and Matsushima is due to the difference in calibration. Our scale is based on a direct comparison of the stellar energy distributions and the models. Terashita and Matsushima folded the model energy distributions with the rather poorly-known UBV filter and sensitivity functions. The zero points of the UBV system, especially of the U filter, are not particularly well known.

Gravities were determined in two ways. From the scans, it was possible to measure the equivalent width of Hβ. A computer program was written to determine the equivalent width of Hβ from the models in the same way as it was measured from the scans, that is, by fixing the 'continuum' as the monochromatic intensity in a certain scanner band (usually ~ 200 Å from line center), folding the model flux distribution with a rectangular instrumental profile (80 or 40 Å wide), and integrating the resulting instrumentally-broadened profile. This procedure insured that models and observations were treated in a self-consistent manner and insured that systematic errors, particularly in the matter of continuum placement, were minimized. Such a procedure is necessary to avoid the ambiguity of defining what is meant by 'the continuum' in stars where most of the Balmer lines overlap.

Equivalent widths can be measured this way to an accuracy of 15%, leading to a random error in the derived value of $\log g$ of about 0.3.

We also determined surface gravities from the profiles determined by Greenstein

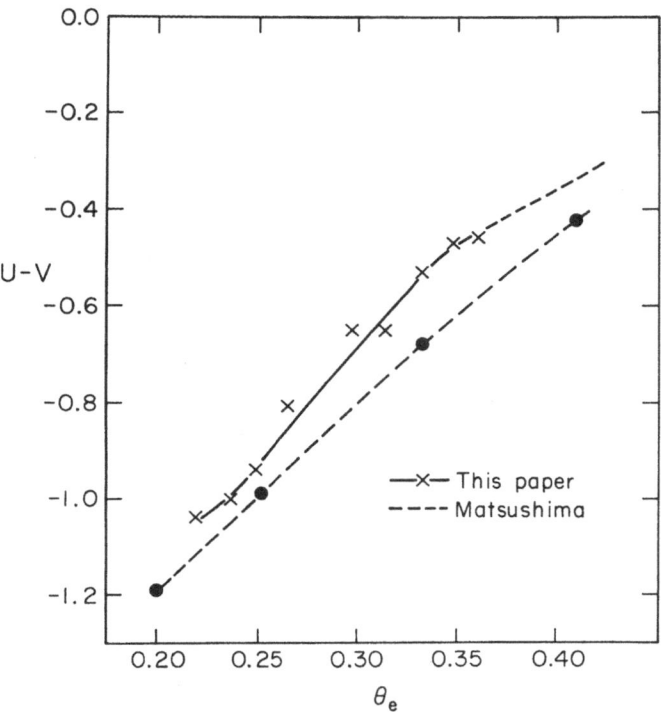

Fig. 7. A plot of the $U - V$ color versus the derived effective temperature $\theta_e = 5040/T_e$ for DA stars.

(1960). Again, in order to treat models and observations in a self-consistent way, the profiles were normalized to zero residual intensity at $\Delta\lambda = 80$ Å. While this procedure neglects the far wings of the line, which certainly extend beyond $\Delta\lambda = 80$ Å, again the ambiguity of continuum placement is eliminated. We estimate the error in surface gravities derived from the Greenstein profiles to be about 0.2 in the logarithm. Both of the above methods give reasonably consistent surface gravities.

For stars with known distances, the radii can be determined from monochromatic magnitudes. From the models, we obtain the relation

$$\log H_{5556} = 1.52 \log (T_e/10^4) - 4.254, \tag{1}$$

where H_{5556}, the flux at 5556 Å, is correct in the range of the models to within 2%. (H_{5556} is normalized so that $\int_0^\infty H_\nu \, d\nu = \sigma T_e^4/4\pi$). From Equation (1) and the calibration of the absolute flux of Vega by Oke and Schild (1970), we derive

$$\log R/R_\odot = - 0.76 \log (T_e/10^4) - \log \pi - 0.2 m_{5556} - 0.50 \tag{2}$$

where m_{5556} is the monochromatic magnitude per unit frequency and π is the parallax. m_{5556} is defined by

$$m_{5556} = - 2.5 \log f_\nu - 48.60 \tag{3}$$

where f_ν is the flux at the frequency corresponding to $\lambda = 5556\,\text{Å}$ in units of ergs $\sec^{-1}\,\text{cm}^{-2}\,\text{Hz}^{-1}$. The above method avoids the uncertainties inherent in the use of bolometric corrections.

6. Helium Abundance

We have calculated He I $\lambda 4471$ profiles for the models using a procedure described earlier (Shipman and Strom, 1970). The model for W 1346 predicts for the 4471 line, $W = 4.7\,\text{Å}$ for a helium abundance of 10% by number and $W = 2.5\,\text{Å}$ for a helium aboundance of 3%. Since $\lambda 4471$ has not been observed in medium-dispersion spectra of W 1346 [Greenstein (private communication) cites an upper limit of 0.6 Å], it is apparent that the atmosphere of W 1346 contains little or no He. We have thus increased all values of $\log g$ deduced from the 10% helium models by 0.12 to correspond to zero helium atmosphere (cf. Matsushima and Terashita, 1969).

7. Masses and Radii

The five stars scanned which have known distances have masses and radii tabulated in Table I. Their position on the mass-radius diagram is shown in Figure 8. It is apparent that the large masses derived by Matsushima and Terashita (1969) were due to too-high surface gravities deduced from too-low electron densities in these models, due to the neglect of pressure ionization in the computation of the ionization equilibrium. The mass-radius curves in Figure 8 are from Hamada and Salpeter (1961). The large uncertainties in the radii are mostly due to the uncertain parallaxes, rather than uncertain temperatures.

Most of the stars seem to have interiors composed of Fe, although the possibility of systematic errors (especially in $\log g$) make this conclusion quite tentative. One star, L1512-34B, which is on the lower sequence, as discussed by Eggen and Greenstein (1965), seems to have a neutron core. In order for this star to fall along the Fe track in the mass-radius diagram, the surface gravity would have to be at least 8.2 even if the radius were as large as the error bars allow. We feel that this can probably be ruled out by the observations. More work is needed on the small-radius white dwarfs of the lower sequence.

It is possible to check the accuracy of our reduction procedures with 40 Eri B, which

TABLE I

Masses and Radii of Five White Dwarfs

Star	T_{eff}	$100\,R/R_\odot$	$\log g$	$M/M_\odot$
40 Eri B	17000	1.27 ± 0.06	7.85	0.42 ± 0.09
W 485	15100	1.46 ± 0.3	7.35	$0.17\ (+0.10, -0.07)$
L1512-34B	13600	0.81 ± 0.3	7.55	$0.085\ (+0.08, -0.04)$
He 3	21300	1.26 ± 0.3	7.60	$0.23\ (\pm 0.05)$
W 1346	20300	1.26 ± 0.2	7.35	$0.125\ (+0.08, -0.04)$

has a mass of 0.43 ± 0.04 $M_\odot$ (Popper, 1954) and a gravitational redshift k of 21 ± 4 km sec^{-1}. Our calculations predict $M = 0.42 \pm 0.09$ $M_\odot$ and $k = 21 \pm 5$ km sec^{-1}. We regard this agreement as excellent. Our temperature is slightly higher than the 16 300 K derived by Matsushima and Terashita (1969) because of the revised calibration of Vega (Oke and Schild, 1970).

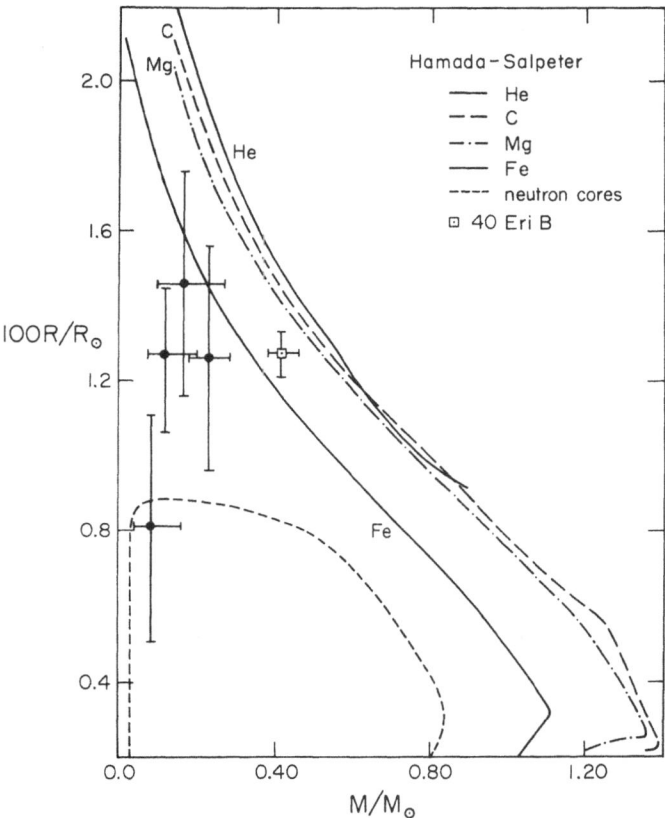

Fig. 8. The mass-radius relation as computed by Hamada and Salpeter (1969). 40 Eri B and several other stars are shown. The error bars reflect mainly the errors in the trigonometric parallaxes.

Furthermore, in deriving the surface gravity of 40 Eri B, we used the photo-electrically-measured Hγ profile of Oke (1963). When normalized to $R_v = 0.0$ at $\Delta\lambda = 80$ Å, the agreement between this profile and that of Greenstein (1960) is quite good. The surface gravities from the Greenstein profile and the Hβ equivalent width are both less by 0.14 in the logarithm than the log$g$ derived from Oke's scan; this is well within the expected error of 0.2 in the logarithm. If, in fact, the surface gravities deduced for the other four stars are slightly small, as this result *might* indicate, the deduced masses will be somewhat greater and the stars will fall nearer the Fe line in the mass-radius diagram.

8. Future Work

Using the $(U-V)-T_e$ relation of Figure 7, we plan to derive masses for as many white dwarfs as we can with known distances. We also plan to scan the Hyades white dwarfs, since their distance is known and their gravitational redshifts have been measured by Greenstein and Trimble (1967).

References

Eggen, O. J. and Greenstein, J. L.: 1965, *Astrophys. J.* **141**, 83.
Eggen, O. J. and Greenstein, J. L.: 1967, *Astrophys. J.* **150**, 927.
Greenstein, J. L.: 1958, *Encyclopedia of Physics* **50**, 161, Springer-Verlag, Berlin.
Greenstein, J. L.: 1960, *Stellar Atmospheres* (ed. by J. L. Greenstein), University of Chicago Press, Chicago, Chapter xix.
Greenstein, J. L. and Trimble, V. L.: 1967, *Astrophys. J.* **149**, 283.
Hamada, T. and Salpeter, E. E.: 1961, *Astrophys. J.* **134**, 683.
Kolesov, A. K.: 1964, *Soviet Astron. AJ* **8**, 185.
Kurucz, R. L.: 1969a, *Astrophys. J.* **156**, 235.
Kurucz, R. L.: 1969b, *Theory and Observation of Normal Stellar Atmospheres* (ed. by O. Gingerich) MIT Press, Cambridge, p. 375.
Matsushima, S. and Terashita, Y.: 1969, *Astrophys. J.* **156**, 219.
Oke, J. B.: 1963, paper presented at the Cleveland meeting of the AAAS.
Oke, J. B.: 1969, *Publ. Astron. Soc. Pacific* **81**, 11.
Oke, J. B. and Schild, R. E.: 1970, *Astrophys. J.* **161**, in press.
Popper, D. M.: 1954, *Astrophys. J.* **120**, 316.
Schatzman, E.: 1958, *White Dwarfs*, North-Holland Publ. Co., Amsterdam.
Shipman, H. L. and Strom, S. E.: 1970, *Astrophys. J.* **159**, 183.
Strom, S. E. and Avrett, E. H.: 1964, *Astrophys. J.* **140**, 1381.
Strom, S. E. and Avrett, E. H.: 1965, *Astrophys. J. Suppl.* **12**, 1.
Terashita, Y. and Matsushima, S.: 1966, *Astrophys. J. Suppl.* **13**, 461.
Terashita, Y. and Matsushima, S.: 1969, *Astrophys. J.* **156**, 203.
Weidemann, V.: 1963, *Z. Astrophys.* **57**, 87.

12. THE NUCLEI OF PLANETARY NEBULAE AS
PROGENITORS OF WHITE DWARFS

C. R. O'DELL

Yerkes Observatory, Williams Bay, Wis., U.S.A.

Stellar evolution is characterized by fast and slow phases. Usually the periods of rapid change are difficult to follow observationally; but, this does not seem to be the case when passing through the planetary nebula stage. Because of their high intrinsic luminosities and easy identification, it is possible to identify and study these objects and their central stars rather completely. It is quite relevant to discuss these objects at a symposium on white dwarfs since the central stars may be in the immediate progenitor stage before white dwarfs. The actual picture of the evolution of the nuclei has changed rather little in the past few years and is the subject of an earlier review article (O'Dell, 1968) to which the reader is referred.

In capsule form, the evolutionary picture is as follows, the systems are discovered soon after ejection of a shell of about 0.3 solar masses and the stars have luminosities about 10^4 solar luminosities. The stars seem to increase in temperture to about 10^5 K at constant luminosity, then to decrease in luminosity very quickly, falling to ten solar luminosities in a total time of about 30000 yr, placing them at the threshold of the hot extension of the white dwarf cooling tracks. It is a suggestive feature of this picture that the luminosities peak close to the radiation pressure boundary for stars of about one solar mass, thus providing an efficient mechanism for mass loss.

There are complications to the above picture that may actually provide fundamental clues to the exact mechanisms occurring. Perhaps most puzzling is the question of multiple ejections of material. This may be evidenced by the double nebula structure of NGC 6543 and NGC 6826. They give every appearance of shell formation at epochs about 30000 yr apart. A photographic search of the brightest dozen other nebulae adds no others to this list. Another such object may be FG Sagittae (Herbig and Boyarchuk, 1968), which is a nebula with a central star of quite cool spectral type. The system is extremely interesting, because the central star is cooling and increasing in luminosity. Herbig and Boyarchuk believe that the absorption spectrum is actually formed in an expanding atmosphere which will become a second shell. It is not at all clear whether the small fraction of double shells should be interpreted as being due to long intervals between ejection or that only a small fraction of the stars create double shells.

Perhaps one of the best methods of testing the model and the several theories would be through the determination of the gravitational redshifts, which yield values of Mass/Radius. Such a program is reasonable because the radial velocity can be obtained from the optically thin nebular shell. The problems due to faintness and weak, broad absorption lines are serious, but not intractable.

Luyten (ed.), White Dwarfs, 77–78. All Rights Reserved.
Copyright © 1971 by the IAU.

The vital statistics for these objects are given in the table below, where the white dwarf data for the rate of formation (x) are due to Weideman (1968). From these numbers, we see that we can account for about 20% of all white dwarfs. The errors in x could allow this fraction to approach 80%. The safe conclusion is that one can account for a substantial fraction of all white dwarfs by stars manifesting the planetary nebula phenomenon.

	Planetaries	*White dwarfs*
Density near sun	1.4×10^{-8} pc^{-3}	10^{-2} pc^{-3}
Formation Rate (x)	0.4×10^{-12} pc^{-3} yr^{-1}	2×10^{-12} pc^{-3} yr^{-1}

References

Herbig, G. H. and Boyarchuk, A. A.: 1968, *IAU Symp.* **34**, p. 383, in D. E. Osterbrock and C. R. O'Dell (eds), *Planetary Nebulae*, Reidel, Dordrecht.
O'Dell, C. R.: 1968, *IAU Symp. op. cit.*, p. 361.
Weideman, V.: 1968, *IAU Symp. op. cit.*, p. 423.

13. THE POLARIZATION OF RADIATION
FROM WHITE DWARFS

J. R. P. ANGEL

Columbia Astrophysics Laboratory, Columbia University, New York, N.Y., U.S.A.

and

J. D. LANDSTREET

University of Western Ontario, London, Ont., Canada

During the course of two observational programs to search for magnetic fields in white dwarfs (Angel and Landstreet, 1970a; Kemp, 1970), it was discovered by Kemp and Swedlund (Kemp *et al.*, 1970b) that the continuum optical radiation from the $\lambda 4135$ type white dwarf Grw + 70° 8247 is circularly polarized. This circular polarization was measured in a broad band from about 4000 to 7000 Å a number of times; in this band the circular polarization does not appear to vary from its mean value of 3.29% with an amplitude of more than about 0.1% (or 3% of the measured mean polarization) on any time scale from 24 seconds to 2 weeks (Angel and Landstreet, 1970b).

The circular polarization has been measured as a function of wavelength with broad-band filters (Angel and Landstreet, 1970b; Gehrels, 1970). It is found that the circular polarization rises sharply from about 0.75% at 3300 Å to a maximum of 3.7% at 4100 Å and then decreases smoothly to about 1.3% at 9400 Å. Further in the infrared, Kemp and Swedlund (1970) have measured the circular polarization to be 8.5 and 15% at 1.15 and 1.25 μ respectively.

It has also been discovered that the light from Grw + 70° 8247 is linearly polarized, and the linear polarization has been measured as a function of wavelength (Angel and Landstreet, 1970b; Gehrels, 1970). The polarization rises from 2.1% at 3300 Å to a maximum of 3.7% at 3800 Å, decreases smoothly to approximately zero at 6400 Å, and then rises to about 2.8% at 9400 Å. Blueward of 6400 Å the position angle of the linear polarization is about 20°, while at the two measured red points at 8200 Å and 9400 Å it is respectively 101° and 148°.

A search for circular polarization in 6 DC stars, one DA, and the peculiar DB star HZ29 has not resulted in the discovery of any other circularly polarized white dwarfs (Angel and Landstreet, 1970b). The probable errors have been in the neighborhood of 0.1%.

It has been argued by Kemp (1970) and shown in laboratory experiments (Kemp *et al.*, 1970a) that electronic radiation from a heated radiator in a strong magnetic field is circularly polarized. It is thought that the continuum circular polarization of Grw + 70° 8247 may be due to the presence of a magnetic field of the order of 10^7 G.

Luyten (ed.), White Dwarfs, 79–80. All Rights Reserved.

References

Angel, J. R. P. and Landstreet, J. D.: 1970a, *Astrophys. J. (Letters)* **160**, L147.
Angel, J. R. P. and Landstreet, J. D.: 1970b, *Astrophys. J. (Letters)* **162**, L61.
Gehrels, T.: 1970, private communication.
Kemp, J. C.: 1970, *Astrophys. J.* **162**, 169.
Kemp, J. C. and Swedlund, J. B.: 1970, *Astrophys. J. (Letters)* **162** L67.
Kemp, J. C., Swedlund, J. B., and Evans, B. D.: 1970a, *Phys. Rev. Letters* **24**, 1211.
Kemp, J. C., Swedlund, J. B., Landstreet, J. D., and Angel, J. R. P.: 1970b, *Astrophys. J. (Letters)* **161**, L77.

14. WHITE DWARF ATMOSPHERES

V. WEIDEMANN

Institut für Theoretische Physik und Sternwarte der Universität, Kiel, Germany

1. Introduction, Surface Gravity

We first consider the general information scheme for the interpretation of observational data (Figure 1). From the relations plotted it is evident that (in going from left to right) this scheme can only be solved if distances are known and if we are able to determine the atmospheric parameters: effective temperature, T_{eff}, surface gravity, g, and chemical composition from observations of colors and spectra – which is the genuine task of the theory of stellar atmospheres.

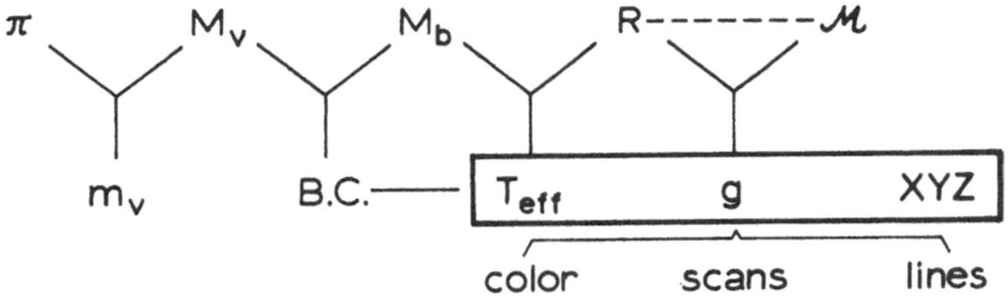

$$\log R/R_\odot = -2\log T_{eff} - 0.2(M_v + \text{B.C.}) + 8.45$$
$$\log \mathcal{M}/\mathcal{M}_\odot = -2\log R/R_\odot + \log g - 4.44$$

Fig. 1. Information scheme displaying relations between observed and derived quantities: π: parallax, m_v and M_v: apparent and absolute visual magnitude, M_b: bolometric magnitude, B.C.: bolometric correction, R and $\mathcal{M}$: radius and mass (broken line indicates mass-radius relation), the box contains the atmospheric parameters: effective temperature, surface gravity g and chemical composition XYZ.

A possibility to solve the scheme in a simpler way is given in the special case only where in addition the mass is known: in this case the problem reduces to a determination of T_{eff} (and bolometric correction B.C.) and g is a derived quantity. If on the other hand a mass-radius relation for degenerate configurations is assumed to hold, (or if reliable gravitational redshifts were available) the scheme can completely be solved in going from right to left, as soon as g and T_{eff} are determined. With this scheme in mind it is easy to bring order into the history of white dwarfs investigations. I shall not do this here but just demonstrate how the *surface gravity* in a typical white dwarf atmosphere can be deduced in the best observed case of 40 Eri B ($\mathcal{M} = (0.43 \pm 0.04)\, \mathcal{M}_\odot, M_v = 11.0$). With a very rough estimate of T_{eff}, $10000 < T_{eff} < 20000$ K ($\Delta \log T_{eff} = \pm 0.15$), B.C. $= (1.0 \pm 0.7)$ we obtain $\log R/R_\odot = -1.85 \pm 0.45$, and, finally $\log g = 7.7 \pm 0.9$.

Luyten (ed.), White Dwarfs, 81–96. All Rights Reserved.

Since this combination of mass and radius fulfills the Chandrasekhar relation for completely degenerate configurations (with $\mu_e = 2$) very well, and since 40 Eri B is a typical white dwarf of the most common spectral type DA we understand that many have taken the validity of the $\mathcal{M} - R$ relation for granted and sometimes used it in an uncritical way. As demonstrated by the example of 40 Eri B the g-determination can be improved if a temperature scale is established. Greenstein's first attempt (1958) was based on calculations of $U - V$ colors with gray atmospheres and estimates of the influence of the Balmer jump on the U band. For 22 white dwarfs with known distances he found the radii to scatter about a factor 2 only around a mean value of $\log R/R_\odot = -1.90$. With the $\mathcal{M} - R$ relation this would imply a range on $\log g$ from about 7 to 8.7. Independent information about the surface gravity can be obtained from line broadening theory, a method which was applied extensively by Weidemann in the case of van Maanen 2 (1960) and the DA white dwarfs (1963) to give $\log g = 8 \pm 0.5$, with a resulting large scatter around the mass-radius relation in those cases where distances were known. In view of the uncertainties in parallaxes and atmospheric parameter determinations it was then only stated that nothing seemed to be in contradiction with a mass-radius relation of $\mu_e = 2$. While this point will be taken up later again, we stress at the moment the fact that the surface gravity in white dwarf atmospheres is 1000 to 10000 times higher than in main sequence stars.

2. DA Atmospheres

For a first estimate on the physical conditions in such high gravity atmospheres with a normal hydrogen-rich composition we can go back to the work of Vitense (1951) who calculated absorption coefficients and the dependence of gas and electron pressure, Balmer jump and other spectral characteristics on effective temperature and gravity all the way up to $\log g = 8$ (see also Unsöld, 1955). The main result is that the high gravity compresses the atmospheres such as to increase the gas pressure by a factor of about 100 as compared to main sequence stars. A correspondingly higher electron pressure is not only responsible for the extreme broadening of the Balmer lines (which overlap and disappear beyond H_ζ) but also for a generally lower ionization degree which shifts the influence of H^- to higher temperatures and causes Balmer jump and lines to weaken below $T_{eff} \approx 12000$ K. We thus understand Figure 2 (Unsöld, 1955; Weidemann, 1966) which is indispensable for a qualitative explanation of the DA distribution in the two-color diagram. Figure 3 summarizes schematically what has been discussed in more detail elsewhere (Weidemann, 1963, 1966): its essence is to consider deviations from a black-body flux distribution (Balmer jump, Balmer lines, H^- maximum) as being reflected in corresponding deviations of the UBV position from the UBV black-body line. Each position is reached in a sequence of steps: black-body to line-free model, (i.e. Balmer jump, H^- shift of the flux to the blue and violet) line-free model to blanketed model (i.e. influence mainly of lines on the stratification and therefore the continuous flux distribution, f.e. by backwarming) and finally corrections for line-blocking effects in the filter bands.

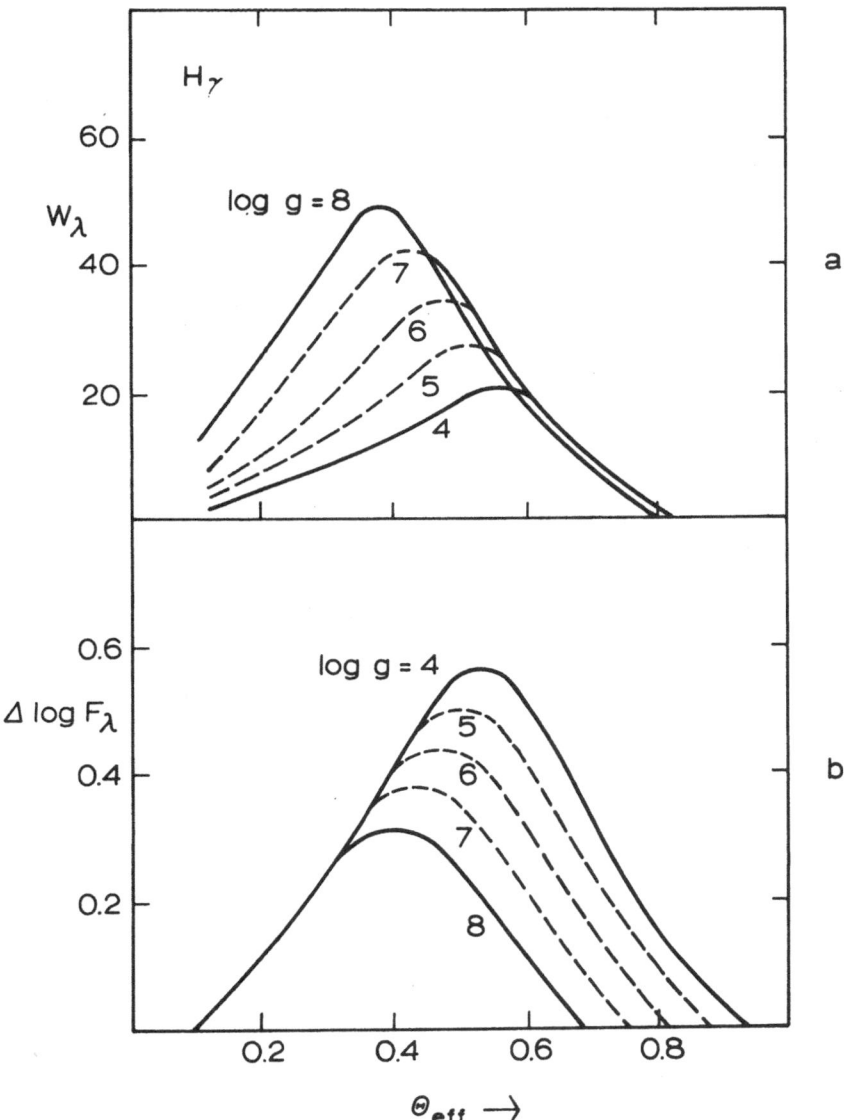

Figs. 2a, 2b.   (a) Equivalent width of Hγ and (b) Balmer jump $\Delta \log F_\lambda$ as a function of $\Theta_{\text{eff}} = 5040/T_{\text{eff}}$, with $\log g$ as parameter. With increasing g the maxima are shifted towards higher T_{eff} due to increasing H^- absorption.

These results were quantitatively confirmed by the model calculations of Matsu-shima and Terashita (1969a), Terashita and Matsushima (1966, 1969) and can be used as an independent method of gravity determination from UBV data (Figure 4) again confirming $\log g \approx 8$ (details will be discussed later). New information is available from intermediate-band photometry by Graham (1970). From Figures 4 and 5 which show essentially line-free positions we may draw two conclusions:

(a) The fact that the DA white dwarfs lie close to the blackbody line in the Johnson system two-color diagram is caused by line-blocking effects rather than a smoothing of the continuum.

(b) The black-body line is not a natural limit for white dwarfs – if there were white

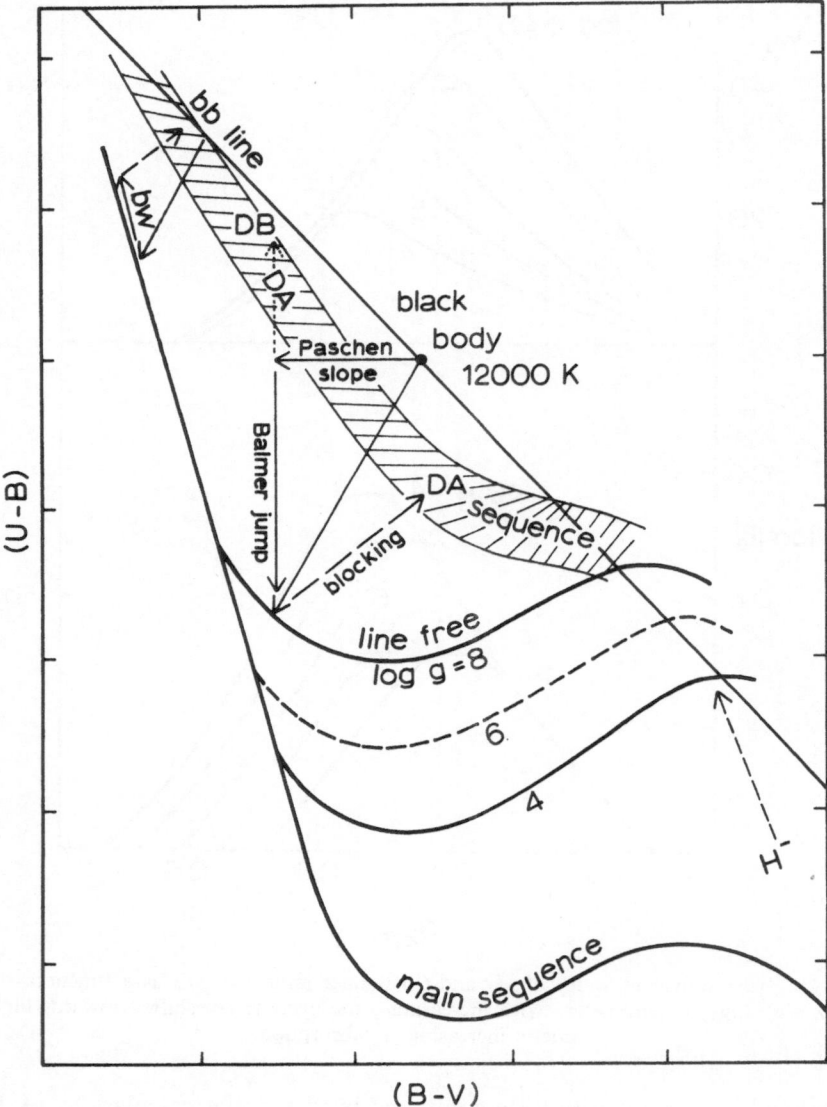

Fig. 3. Schematical behavior of DA white dwarfs in the UBV diagram: the deviations from a black-body flux-distribution (Paschen slope, Balmer jump) add up to give the line-free positions, from which a blocking vector (broken line) reaches the observed position. Upper left: a backwarming vector (bw) operates along the line-free position curve and causes the star to appear hotter. Lower right: H⁻absorption, strongest in the red, shifts the flux to the blue and the ultraviolet (broken arrow). Dotted arrow points to DB position (see text on p. 94).

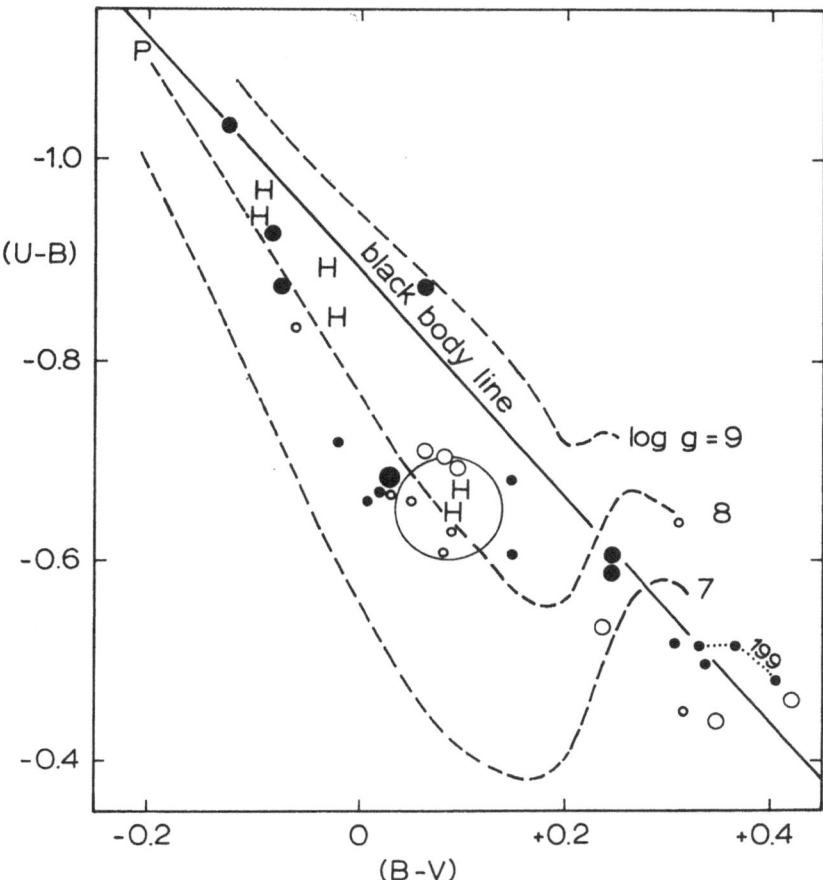

Fig. 4. UBV diagram with black-body line, theoretical positions for $\log g = 7$, 8 and 9, according to Terashita and Matsushima (1969) (broken lines), and observed DA white dwarfs with known distances (Eggen, 1969): ●: 40 Eri B, ⬤: Eggen Table I, ○: Table II, •: Table IV, ○: Table VI, H: Hyades, P: Pleiades. Dotted line connects different positions for EG 199. Circle indicates maximum error of 0.05 mag.

dwarfs with $\log g$ about 9, we would according to Figure 4 expect them above this line (where e.g. many blue high galactic latitude objects are found).

In a similar way the behavior of the Balmer lines can be understood. Whereas the equivalent width reaches a maximum at $T_{\text{eff}} = 12000$ K for $\log g = 8$ (Figure 2) due to the increased H$^-$ opacity, the central intensity decreases all the way from the hottest to the coolest DA stars. At the cool end of the sequence the Balmer lines thus appear sharp (DAs, DAss), at the hotter shallow (DAwk). The variation of the profile shape of Hγ as a function of both T_{eff} and g has been determined first semi-empirically by Weidemann (1963), then calculated by Matsushima and Terashita (1966), and with a revised Stark broadening theory by Terashita and Matsushima (1969). Whereas I used these profiles in 1963 to derive (essentially relative) T_{eff} and g-values for 22 DA

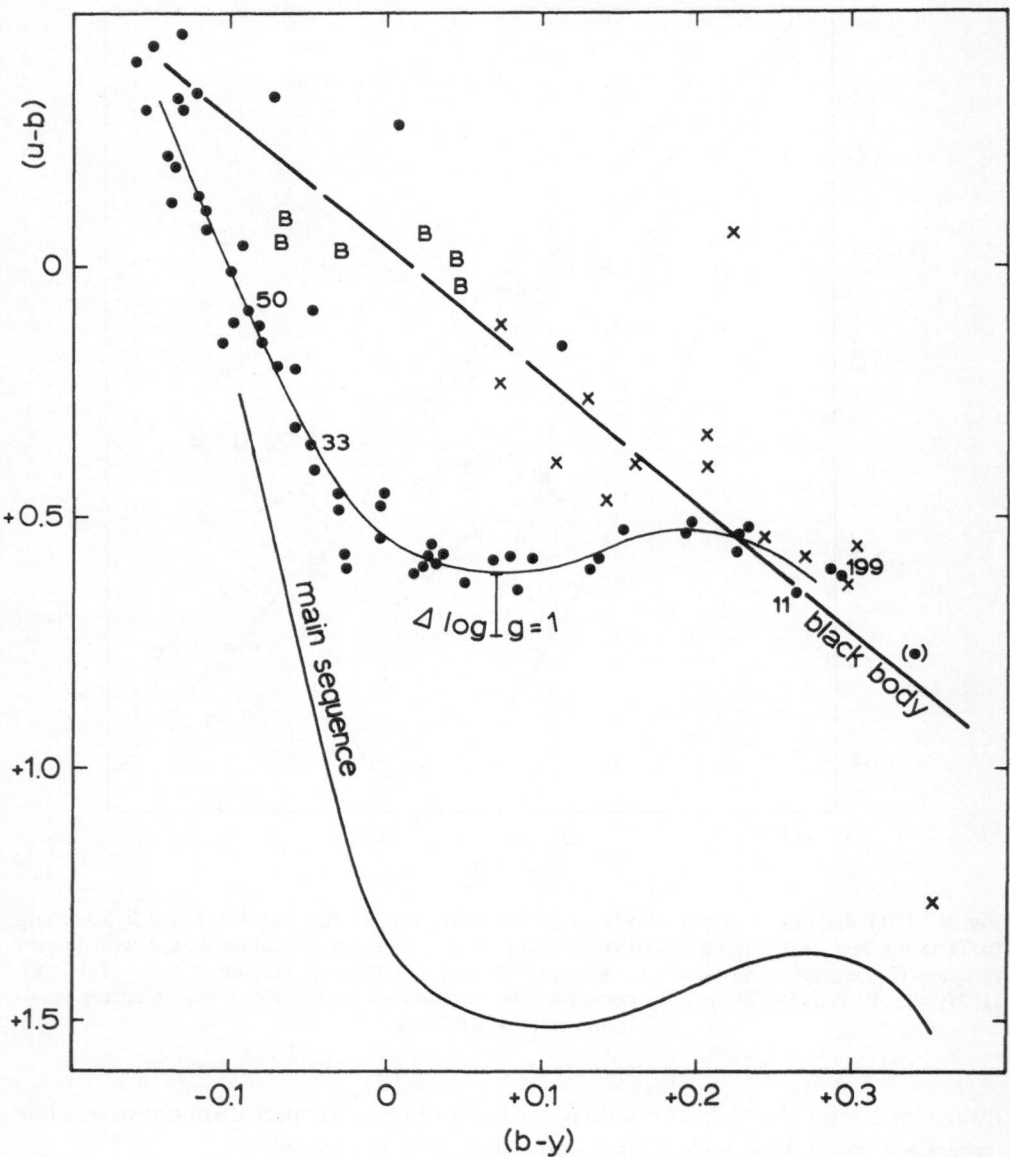

Fig. 5.   Two-color diagram in the Strömgren system, according to Graham (1970). Dots: DA (and
DO) white dwarfs, B: DB, crosses: other non-DA white dwarfs (●): EG 87, composite, not reliable.
Mean DA sequence outlined. Bar indicates theoretical shift for a gravity change (cf. Figure 3
and text, p. 88).

stars, Matsushima and Terashita preferred to determine these parameters from the
two-color diagram (Figure 4) since their (purely theoretical) model-atmosphere
approach did not reproduce the observed profiles well enough. Considerable differences in gravity determinations between Matsushima and Terashita (1966) and

Terashita and Matsushima (1969) are due to the fact that different line-broadening theories have been used. Since the broadening theory is still under discussion, it seems that even now line-profile data should be used only differentially for the determination of atmospheric parameters.

As far as the models themselves are concerned, the full capacity of computational facilities has been applied in the second half of the past decade by Matsushima and

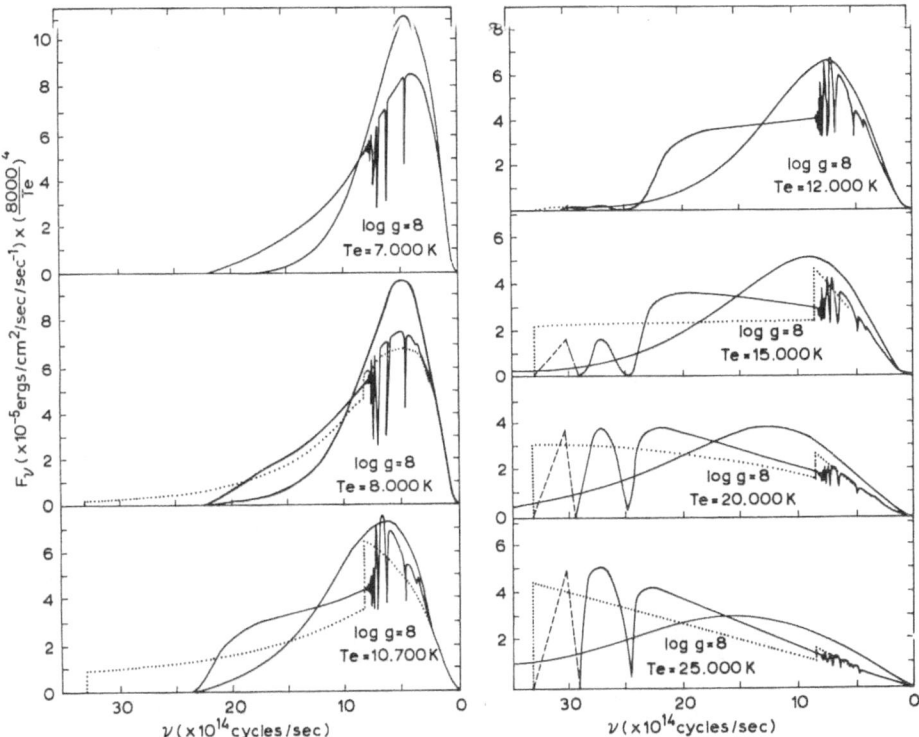

Fig. 6.   Flux distributions for DA model-atmospheres, $\log g = 8$, according to Terashita and Matsushima (1969). Dotted: non-blanketed models. Black-body distribution for comparison.

Terashita. Although the first approach (1966) in which radiative flux-constant unblanketed models were calculated was only partly successful (Weidemann, 1968), the newer models in which the blanketing by hydrogen lines was fully taken into account (Terashita and Matsushima, 1969) should constitute as real step ahead. Figure 6 (reproduced from TM, 1969) clearly demonstrates how Lyman absorption causes backwarming (especially for the hotter models) and how the emergent flux in the visual regions changes. For the first time bolometric corrections were derived and a new temperature scale was established. Terashita and Matsushima were thus able to solve our information scheme (Figure 1) with higher accuracy than before and determined radii and masses for 30 DA stars with known luminosities. They finally went on (MT, 1969b) to admit variations in the ratio of hydrogen to helium abundances, and considered the special case of 40 Eri B (MT, 1969c). With all other data known a value of 0.90 for the fraction of hydrogen by number was derived. Although the absolute data given should not be considered final in view of the fact that metals were not included as opacity sources (see below, p. 92), and in view of discrepancies to be discussed later these papers demonstrate clearly the power of computer methods in the case of white dwarf atmospheres. However, doubts arise if we next turn to the question of the significance of the results. Radii and masses as determined by Terashita and Matsushima

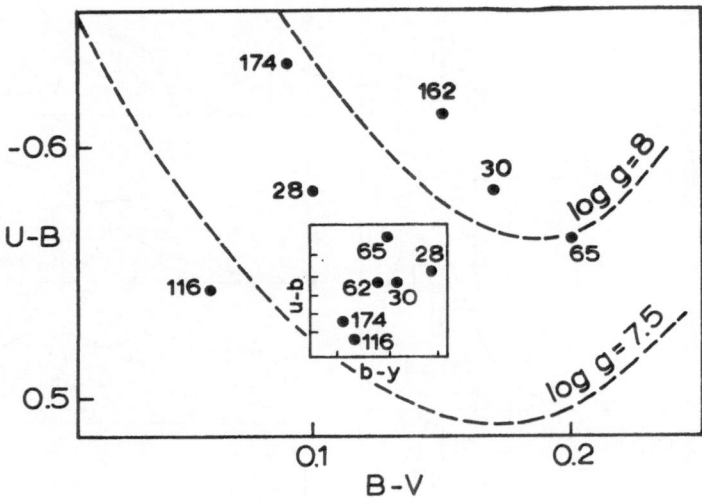

Fig. 7. Comparison of UBV and uby data for six DA stars. Broken lines in the UBV diagram indicate theoretical lines for constant g according to Terashita and Matsushima (1969). Insert is adjusted in scale such as to cover corresponding intervals in $(u-b)$ (scale 0.01 mag) and $(b-y)$ (scale 0.01 mag). Graham error (1969) is ± 0.01 mag in $(b-y)$ and ± 0.015 in $(u-b)$.

(1969) do show a very large scatter around the theoretical mass-radius relation (Figure 9). Since the radii were derived from absolute visual magnitudes via T_{eff} and masses from radii with $\log g$, and since both, T_{eff} and $\log g$ were read out from the position of individual stars in the two-color diagram (Figure 4) the reliability of the results in this approach depends on both the significance of the scatter in the color-magnitude diagram and the UBV diagram.

In this respect the narrowness of the DA sequence in Graham's two-color diagram (Figure 5) teaches us that most of the scatter in Figure 4 cannot be significant. This was evident in some cases for which UBV has been determined repeatedly – e.g. for LP 9-231 (EG 199) (van Altena, 1966; Luyten, 1967; Eggen and Greenstein, 1967; Eggen, 1969) differences up to 0.08 mag in $(B-V)$ or 0.12 mag in $(U-V)$ occur in the published data. But more convincing is the following consideration. We first extract from the predicted locations for line-free flux distributions (Figure 3) that the g-dependence in the Strömgren system is largest for the cooler DA stars with $0 < (b-y) < +0.2$. From Figure 3 and Figure 5 – in which a bar indicates the shift of the line-free sequence for $\Delta \log g = -1$, calculated with TM fluxes (1969) – we then see that this part of the Graham sequence implies for the observed DA stars $\log g = $ const. to within ± 0.15. If we furthermore check if there is any correlation between relative positions in the Johnson and the Strömgren two-color diagrams we obtain a negative result. Figure 7 enlarges a part of both diagrams in comparable scales and shows how a group of 6 DA white dwarfs which in the Strömgren system have nearly identical positions are 'blown up' in the Johnson system. Again corrections up to 0.08 mag in UBV colors must be invoked in order to obtain a comparable scatter in $\log g$.

Finally I plotted a histogram (Figure 8) which shows how the scatter in g-values derived is reduced by changing from Johnson to Strömgren colors. The obvious conclusion is that the DA stars do in reality form a rather narrow sequence and that the g-scatter derived from the UBV diagram is not significant. It follows that the DA

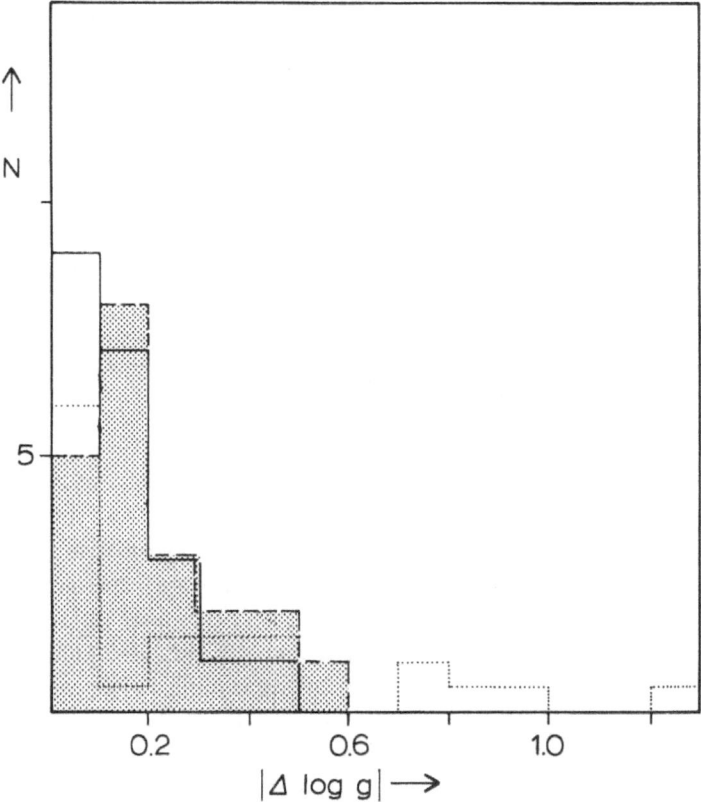

Fig. 8. Distribution of DA white dwarfs around mean sequence of constant gravity. Full line: on using data from Graham's two-color diagram (Strömgren system) (1970); dashed line, shaded area: from UBV diagram (Figure 4); dotted: from M_v-(U − V) diagram (Figure 10).

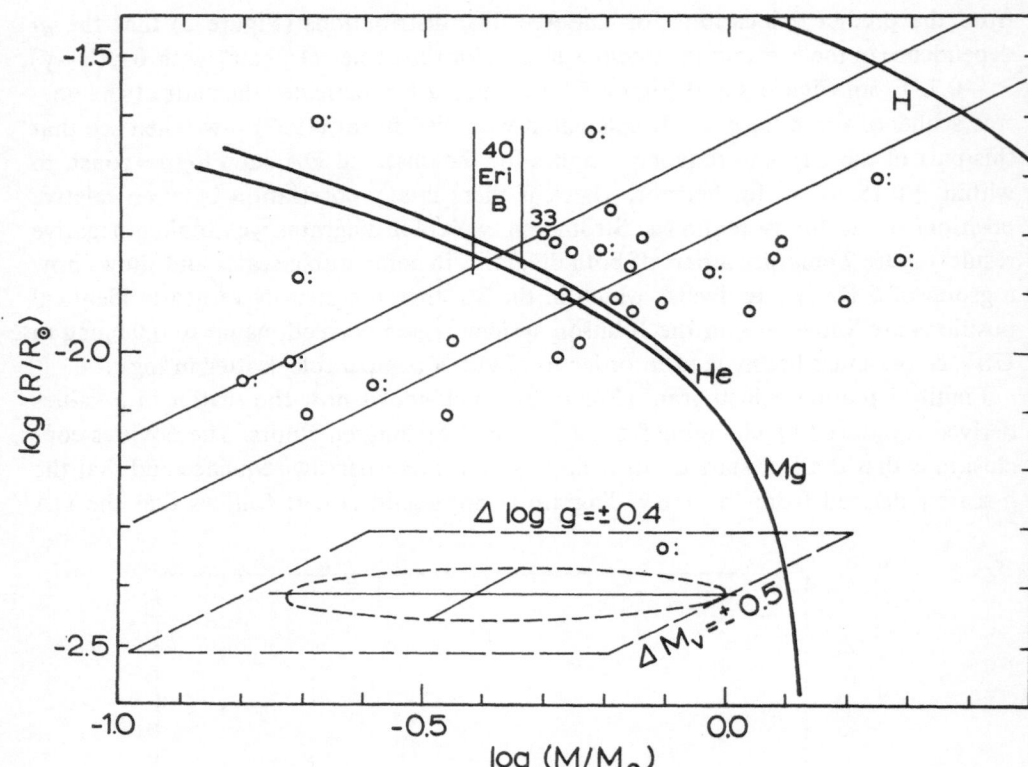

Fig. 9. Mass-radius diagram, as given by Terashita and Matsushima (1969) with additions: diagonal strip indicating $\log g =$ const. ± 0.15, to which white dwarfs should be confined if Graham's data and interpretation given here are accepted; vertical strip: mass range of 40 Eri B (EG 33); lower part insert: error ellipse corresponding to error circle in UBV-diagram of Figure 4, dashed lines indicating total error area including luminosity error. Notice same general size and direction of derived white dwarf data cloud. Colons mark stars for which either distance or spectral type is called uncertain in the literature.

stars in the mass-radius diagram of Terashita and Matsushima (1969) should be confined to a narrow strip around $\log g =$ const. (Figure 9). The remaining scatter is then entirely due to scatter in radii. From the shape of the distribution alone it can be concluded that this scatter too has probably no physical meaning since if real it would imply a mass-radius relation which runs just diagonal to the Chandrasekhar relation for degenerate stars. Although it cannot be excluded that there are some objects with extended hydrogen shells and corresponding larger radii (Hamada and Salpeter, 1961) it is more probable that the entire scatter is due to errors in UBV colors and distance determinations. (See error area outlined in Figure 9.) In order to check this further we consider finally the color-magnitude diagram (Figure 10). It shows that the stars with better parallaxes (large dots, Eggen, 1969 and 1970) do not contradict a theoretical mean relation for completely degenerate configurations of $\log g =$ const. (broken lines, according to Matsushima and Terashita, 1969a). [In the case of

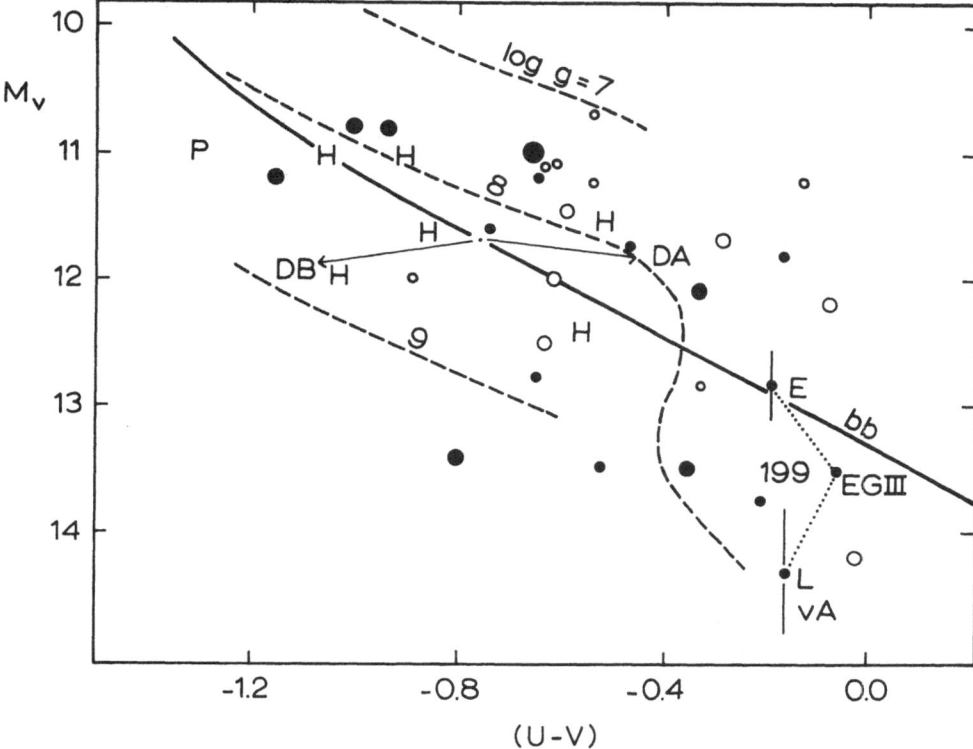

Fig. 10. Color-magnitude diagram for DA stars. Symbols as explained in legend to Figure 4. Dashed lines: theoretical lines for constant radii (and corresponding gravities for fully degenerate configurations) according to Matsushima and Terashita (1969a) with black-body line for $\log g = 8$, $\log R/R_{\odot} = -1.91$, added. Arrows arising from black body at 12000 K demonstrate differences of theoretical DA and DB positions at equal $T_{\rm eff}$ (cf. Figure 3 and text on p. 94).

Lower right: different data extracted from the literature for EG 199: EG III (Eggen and Greenstein, 1967), L (Luyten, 1967), vA (van Altena, 1966), E (Eggen, 1969) indicate extreme uncertainties of M_v determinations in some cases.

EG 165, far to the left, $U-V$ is probably in error since it fits the sequence of the other objects in the M_I $(R-I)$ diagram of Eggen (1970)]. The large discrepancies of individual parallax determinations are demonstrated in the (admittedly extreme) case of EG 199.

Correspondingly the scatter of $\log g$ if derived from this diagram via the $M-R$-relation would be largest (see histogram Figure 8, dotted line). We thus conclude that there is strong evidence that the DA white dwarfs are confined to essentially constant surface gravity and do obey the mass-radius relation. The absolute figures depend somewhat on the scale calibrations: my best estimate is at present $\log g = 7.90 \pm 0.15$ corresponding to masses between 0.42 and 0.60 solar masses. In view of the importance of this conclusion we must now ask how far it depends on uncertainties or imperfections of atmospheric theory. A few facts must be mentioned in this connection

(1) From a comparison between the shape of the theoretical curves for $\log g = $ const. in both two-color diagrams and the observed DA distributions in Figure 4 and Figure 5 it is evident that Terashita and Matsushima's calculations predict a too pronounced S-shape at the cool end of the DA sequence. This may also be concluded by a comparison of the theoretical color-magnitude relation (Figure 10) with the observed monotonic decrease of W_λ (Hγ) as a function of $(U-V)$ (see e.g. Greenstein, Figure 3, 1969). Since the turn-up of the DA sequence at the cool end is caused by H^- (Figure 3) one probably has to invoke a reduction of H^- relatively to other opacity sources. At lower temperature the atmospheres become extremely neutral with $\log(P_g/P_e) \approx 4$. Metals and molecule formation cannot be neglected and may be responsible for the discrepancies mentioned. Unfortunately we have no observational hint as yet on metal abundances in DA stars. Assuming it to be normal we do expect extreme blanketing and blocking effect on coming down to solar temperature (see Weidemann, 1966). A study is underway at Kiel to get some quantitative results.

(2) As for the reliability of the TM temperature scale at higher temperatures, around 15000 K, doubts remain since backwarming effects caused by metal absorption in the UV should become important. For main sequence stars, Davis and Webb (1970) have pointed out that Mihalas hydrogen-line-blanketed models – which correspond in the degree of sophistication to those of Terashita and Matsushima – yield upper limits to effective temperatures only, since the space-observed UV depressions are larger, whereas Adams and Morton (1968) have shown that metal line blanketing causes backwarming amounting to about 1000 degrees at T_{eff} around 16000 K.

For white dwarfs with normal metal abundances UV line blanketing should be even stronger due to increased pressure broadening.

Indeed, Graham's most accurate data (1970) give some indication that the temperature scale should be stretched between 40 Eri B (EG 33) – for which Matsushima and Terashita, 1969c, give $T_{eff} = (15300 \pm 380)$ K – and He 3 (EG 50),: a dip in the m_1 vs. $b-y$, correspondingly stronger in the c_1 vs. $u-y$ diagram and a plateau between both stars in both the M_v vs. $u-y$ and $U-V$ color magnitude diagrams point all into this direction. The sudden steepening of the visual flux distribution necessary to explain these phenomena could naturally arise from backwarming which becomes strong when the flux maximum shifts into the UV region below 3000 Å.

3. Non-DA Atmospheres

It is evident from the spectra and especially from relative positions in the two-color diagrams that the atmospheric composition for the non-DA spectral types (DB, DC, λ4670) is different. (Greenstein, 1958, 1960, Weidemann, 1968).

Knowing with certainty that the interior of degenerate stars must be hydrogen-free we may be surprised to find hydrogen in the atmospheres at all. The fact that we observe hydrogen spectra in the DA's does not imply that H is the most abundant atmospheric constituent. First estimates (Weidemann, 1963) and subsequent calculations by Nariai and Klinglesmith for pure H/He mixtures (1969) and for models with variable H/He ratios including metals by Bues (1970) do show clearly that due to the enormous efficiency of H absorption the visual spectra of stars with $\log g = 8$, $T_{\text{eff}} = 15\,000$ K remain hydrogen dominated even down to H/He ratios of 10^{-3}.

Vice versa Miss Bues was able to show that the complete absence of hydrogen lines in the HeI-line DB spectra forces the atmospheric hydrogen fraction down to about 10^{-5}. This extremely low value was confirmed by a detailed study of the helium line profiles, whereas from colors alone an upper limit for H/He of 10^{-4} was derived which is independent of metal content. The fact that in DB atmospheres hydrogen is completely absent is certainly of evolutionary significance. Since there is further indication that heavy elements are also underabundant (about a factor of 10 to 100 as compared with solar composition) we have to deal with atmospheres which consist of helium to a degree of purity never encountered in other astronomical objects.

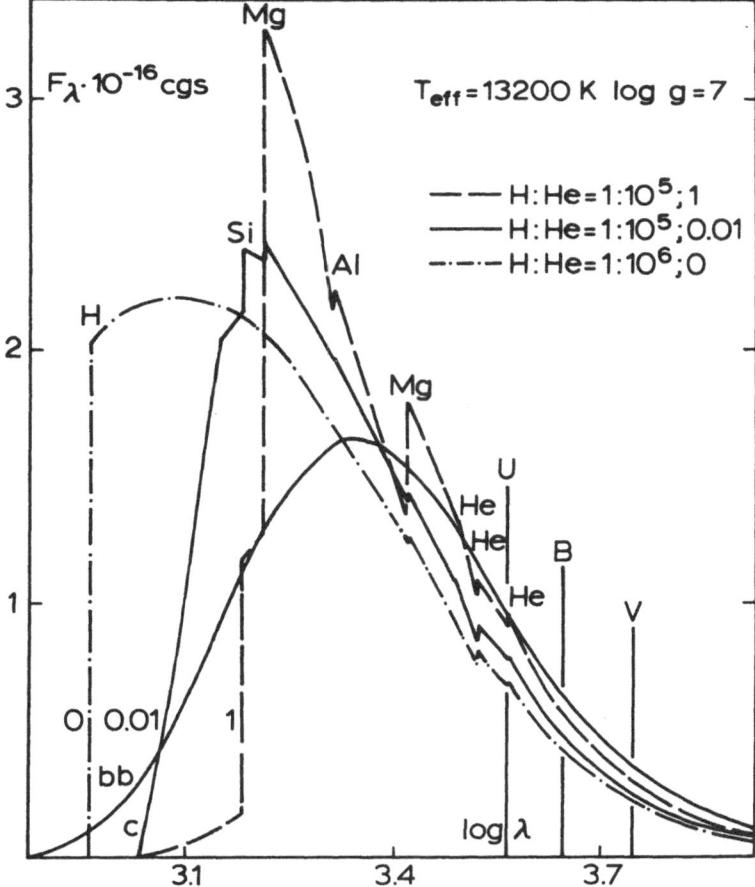

Fig. 11.   Flux distribution for a DB model, according to Bues (1970), demonstrating the importance of metal absorption in the UV and its influence on the visual emergent flux (U, B, and V marked). Different mixtures as indicated: 1: solar metal abundance, 0.01: 1 % solar metal abundance. bb: black-body distribution.

Although the metal content may be low it cannot be neglected in the construction of DB model atmospheres. Figure 11 demonstrates how the important absorption edges of Mg, Si and C in the far UV cause the flux distribution in the visual region to steepen beyond that of the corresponding black body. This effect may in reality be even larger since the models do not include metal line blanketing which again – as in the case of the DA's – should increase the backwarming. This has immediate consequences for the temperature scale of the DB's: they do appear hotter than they actually are. Since for DB stars a Paschen like slope continues from the V via the B to the U band the shifting-vector from a black body to a line-free star in the two-color diagram should point along the black-body line to the upper left of Figure 3 – a fact which again has been confirmed by Miss Bues calculations. We thus understand her result that the DB stars are actually much cooler than anticipated from the UBV position relative to the DA's or black-bodies. Indeed the relative shift between a DA and a DB star of equal effective temperature $(12000 \text{ K}, \log g = 8)$ amounts to $\Delta (U - V) = 0.6$ (Figure 10). This demonstrates which extreme care should be taken in interpreting relative positions in color-magnitude diagrams as being due to relative changes in radii (or, if completely degenerate configurations are assumed, masses): the fact that most non-DA stars are located below the bulk of the DA's in the $M_v - (U - V)$ diagram essentially reflects the absence of a Balmer discontinuity and does not give any information about differences in masses and radii between DA and non-DA white dwarfs. Since in the case of the comparatively rare DB's there are no close objects and no reliable distance determinations and since furthermore the spectra depend only little on gravity we are unfortunately not able to check if the $M - R$ relation is fulfilled. Yet it can again be stated that nothing is contradicted. (For a more detailed discussion of the DB stars I refer to Miss Bues' publication.) With regard to the atmospheric conditions we add that with $\log P_g \approx 7.5$ and $\log P_e \approx 4$ we have extremely neutral matter, with electrons provided by the metals and He^- as the most important opacity source in the visual region.

The trend towards neutrality and higher densities continues with decreasing effective temperatures. Calculations for cooler model atmospheres are at present being made in view of a possible explanation of other non-DA white dwarfs which are located below the DB group in the two-color diagrams, more or less closely along the black-body line. Whereas Miss Bues (1970) checked on the visibility of the H and K lines, Kumar and Doyle (1970) state that the CI white dwarf EG 182 and the $\lambda 4670$ stars are definitely rich in carbon. Since this is not the case for the DB's it seems that the minimum hypothesis of *one* second extremely hydrogen-deficient non-DA sequence of cooling white dwarfs (Weidemann, 1968) connecting DB, DC, $\lambda 4670$, and possibly DG stars cannot longer be upheld. On the other hand it might still be possible that cooling DB stars become DC objects. With regard to the DC stars which show only continuous spectra Wickramasinghe and Strittmatter (1970a) checked a suggestion by Ostriker and Bodenheimer (1968) that these might be normal white dwarfs for which extreme rotation has washed out the lines. Their calculations show that although this could be the case for absorption lines of other elements it cannot cause the hydrogen lines to disappear. Investigations for extreme hydrogen-poor atmospheres at lower temperatures become hampered by an increased tendency to molecule or quasi-molecule formation which changes both the equilibrium composition and the opacity sources.

The fact that molecule formation plays an important role at solar temperatures had already been demonstrated in the analysis of the atmosphere of van Maanen 2, the only representative of spectral type DG (Weidemann, 1960). In this case it was found that for the most probable model with $\log g = 8$ and solar effective temperature hydrogen atoms and molecules should be present in equal amounts, whereas the total hydrogen abundance compared to helium had to be reduced at least ten times in order to yield the high transparency necessary for the explanation of the extreme pressure broadening of the observed iron lines. Another unexpected result was the conclusion that metals should be underabundant by a factor of 10000, slightly dependent on the assumed temperature. The atmospheric pressure was estimated to be around 200 atm ($\log P_g = 9.3$). From recent observations in the infrared (Eggen, 1970) and from Strömgren colors (Graham, 1969) we now find that van Maanen 2 should be slightly hotter than the sun with $T_{eff} \approx 6700$ K and a corresponding reduction of molecule formation. However even in this case the atmosphere remains extremely neutral with $\log (P_g/P_e)$ about 10^7. With a smaller radius $\log g$ increases to about 8.5 and the mass to $0.9\ M_\odot$ if the mass-radius relation is assumed to hold. Under such conditions the scale height of the atmosphere shrinks to 4 m only (as compared to 200 km in the sun). Is is evident that care must be taken to apply the usual theory, f.e. of pressure broadening. The fact that the Fe lines in vMa 2 are symmetrically broadened on the other hand gives an independent hint that helium with its low polarizibility is the main broadening agent.

Studies on the possibility of convection in the outer nondegenerate envelopes of hydrogen-poor white dwarfs have been carried out by Böhm (1968, 1969) van Horn (1970) and Grenfell and Böhm (1970). Although these papers aim mainly at the question of energy transport from the interior to the surface (in connection with the cooling theory) it is interesting in our context to note that the upper boundary of a He^- convection zone for a set of models with van Maanen 2's composition and effective temperatures between 5000 and 14000 K (with absorption of H, H^-, He^-, C, N, O, C^-, O^-, Mg$\textsc{i}$ and Si$\textsc{i}$, as well as H_2-molecule formation included) reaches its maximum for T_{eff} about 6000 K at an optical depth of about 0.5. As in the case of DA and hot pure-helium atmospheres which have been studied by Wickramasinghe and Stritt-matter (1970b) convection turns out to be practically unimportant for the interpretation of the visual spectra: whereas the equivalent widths of Hγ and He$\textsc{i}$ $\lambda4472$ are increased by about 10% – well below observational and theoretical uncertainties – the colors do even change less (0.01 mag in $U-B$ and $B-V$). This insensitivity of colors and equivalent widths to slight changes of temperature stratification has also been demonstrated for DA atmospheres with varying H/He ratios by Matsushima and Terashita (1969b).

It thus appears that in the construction of real white dwarf atmospheres convection is less important than the inclusion of other opacity sources and the consideration of blanketing effects for all ranges of temperatures and composition. In view of the scarcity of information it is hoped that new observations in a broader spectral range (e.g. multichannel data) might help us to narrow down the range of possibilities with which we certainly will have to live for some time to come.

References

Adams, T. F. and Morton, D. C.: 1968, *Astrophys. J.* **152**, 195.
Bues, I.: 1970, *Astron. Astrophys.* **7**, 91.
Böhm, K. H.: 1968, *Astrophys. Space Sci.* **2**, 375.
Böhm, K. H.: 1969, in *Low Luminosity Stars* (ed. by S. S. Kumar), Gordon and Breach, New York, p. 393.
Davis, J. and Webb, R. J.: 1970, *Astrophys. J.* **159**, 551.
Eggen, O.: 1969, *Astrophys. J.* **157**, 287.
Eggen, O.: 1970, *Astrophys. J.* **159**, 945.
Eggen, O. and Greenstein, J. L.: 1967, *Astrophys. J.* **150**, 927.
Greenstein, J. L.: 1958, in *Encyclopedia of Physics* **50**, Springer-Verlag, Berlin, p. 161.
Greenstein, J. L.: 1960, in *Stars and Stellar Systems* **6** (ed. by J. L. Greenstein), University Chicago Press, Chicago, p. 676.
Greenstein, J. L.: 1969, *Astrophys. J.* **158**, 281.
Graham, J. A.: 1969, in *Low Luminosity Stars* (ed. by S. S. Kumar), Gordon and Breach, New York, p. 139.
Graham, J. A.: 1970, *White Dwarf Photometry: Extended List*, private communication.
Grenfell, T. C. and Böhm, K. H.: 1970, *Astrophys. J.* **161**,1183.
Hamada, T. and Salpeter, E. E.: 1961, *Astrophys. J.* **134**, 683.
Luyten, W. J.: 1967, *Publ. Astron. Obs. Minnesota* **3**, Nr. 19.
Matsushima, S.: 1969, *Astrophys. J.* **158**, 1137.
Matsushima, S. and Terashita, Y.: 1969a, in *Low Luminosity Stars* (ed. by S. S. Kumar), Gordon and Breach, New York, p. 315.
Matsushima, S. and Terashita, Y.: 1969b, *Astrophys. J.* **156**, 183.
Matsushima, S. and Terashita, Y.: 1969c, *Astrophys. J.* **156**, 219.
Nariai, K. and Klinglesmith, D. A.: 1969, in *Theory and Observation of Normal Stellar Atmospheres, Proc. 3rd Harvard Conf.* (ed. by O. Gingerich), MIT Press, Cambridge, Mass. and London, p. 315.
Ostriker, J. P. and Bodenheimer, P.: 1968, *Astrophys. J.* **151**, 1089.
Terashita, Y. and Matsushima, S.: 1966, *Astrophys. J. Suppl.* **13**, 461.
Terashita, Y. and Matsushima, S.: 1969, *Astrophys. J.* **156**, 203.
Unsöld, A.: 1955, in *Physik der Sternatmosphären*, 2nd ed., Springer-Verlag, Berlin, p. 202ff and p. 479.
Van Altena, W. F.: 1966, *Publ. Astron. Soc. Pacific* **78**, 345.
Van Horn, H. M.: 1970, *Astrophys. J.* **160**, L53.
Vitense, E.: 1951, *Z. Astrophys.* **29**, 73.
Weidemann, V.: 1960, *Astrophys. J.* **131**, 638.
Weidemann, V.: 1963, *Z. Astrophys.* **57**, 87.
Weidemann, V.: 1966, *J. Q. Spectr. Radiative Transfer* **6**, 691.
Weidemann, V.: 1968, *Ann. Rev. Astron. Astrophys.* **6**, 351.
Wickramasinghe, D. T. and Strittmatter, P. A.: 1970a, *Monthly Notices Roy. Astron. Soc.* **147**, 123.
Wickramasinghe, D. T. and Strittmatter, P. A.: 1970b, *Monthly Notices Roy. Astron. Soc.* **150**, 435.

15. COOLING OF WHITE DWARFS

H. M. VAN HORN

Dept. of Physics and Astronomy and C. E. Kenneth Mees Observatory
University of Rochester, Rochester, N.Y., U.S.A.

Abstract. A knowledge of the precise relationship between the age and luminosity of a white dwarf can in principle be used to determine the compositions of the white dwarfs in galactic clusters. To this end the assumptions in Mestel's theory of white dwarf cooling are critically reviewed, and the results of recent work aimed at relaxing these restrictions are briefly summarized. It is concluded that on the basis of current knowledge an accuracy of the order of 10 or 20% in the age-luminosity relation should be attainable.

1. Introduction

The theory that the luminosity of a white dwarf is derived from the thermal energy content of the ions in its interior was proposed by Mestel (1952). In this paper Mestel establishes two important points: (1) that the relation between the luminosity L of a white dwarf and the cooling time t is $t \propto L^{-5/7}$, for Kramers' law opacity in the envelope, and (2) that the theoretical age of a white dwarf is inversely proportional to the mean atomic weight A of the ions in its interior. For faint white dwarfs such as van Maanen 2 this leads to the conclusion that A must be of the order of 10 in order that the age of the star be less than the age of the Galaxy. The approximations in the theory, however, preclude the possibility of distinguishing between $A=4$ and $A=24$. Consequently, it has not yet been possible to determine from the observational data what phase (or phases) of nuclear burning occur immediately prior to evolution to the white dwarf stage.

It is the aim of this review to show that recent advances have brought the theory of white dwarf evolution to the stage where theoretical ages with an uncertainty of the order of tens of percents, rather than factors of two, are now possible. With accuracies of this sort, comparison of the theoretical age-luminosity relation with the known ages of the galactic clusters can provide reliable determinations of the mean chemical compositions of the cluster white dwarfs. Similarly, the shape of the observed white dwarf luminosity function, which is simply related to the age-luminosity relation and is primarily affected by the thermal structure of the envelope, may lead to important new conclusions about the outer layers of the white dwarfs.

Since a number of excellent reviews (Schwarzschild, 1958, Chapter 7; Mestel, 1965; Weidemann, 1968) have dealt with various aspects of Mestel's original theory, I shall mainly restrict myself to discussions of the newer developments in the theory of white dwarf cooling. It is convenient for this purpose to begin with a critical summary of the Mestel theory, with particular emphasis on those approximations that have been shown to lead to uncertainties of the order of a factor of two in the theoretical age-luminosity relation.

Luyten (ed.), White Dwarfs, 97–115. All Rights Reserved.
Copyright © 1971 by the IAU.

2. Critical Review of Mestel's Theory

The luminosity of a star is in general given by (see, e.g., Hayashi *et al.*, 1962, p. 39)

$$
L = \int_0^M \left(\varepsilon - T \frac{\partial s}{\partial t} \right) dM_r,
\tag{1}
$$

where ε is the rate of energy generation (or loss) per unit mass by nuclear processes, $T \partial s/\partial t$ is the time rate of change of the heat content per unit mass, and M_r is the mass interior to radius r. Since s, the entropy per gram, can be regarded as a function of the temperature T and density ϱ of the stellar matter, we have

$$
T \frac{\partial s}{\partial t} = T \left[\frac{\partial s}{\partial T} \bigg|_\varrho \frac{\partial T}{\partial t} + \frac{\partial s}{\partial \varrho} \bigg|_T \frac{\partial \varrho}{\partial t} \right] = C_v \frac{\partial T}{\partial t} - \frac{T}{\varrho^2} \frac{\partial P}{\partial T} \bigg|_\varrho \frac{\partial \varrho}{\partial t}.
\tag{2}
$$

If nuclear (and neutrino) processes are neglected, following Mestel, (Approximation 1), and if the energy released by residual gravitation contraction is ignored ($\partial \varrho/\partial t = 0$: Approximation 2), the luminosity of a white dwarf is directly proportional to the time rate of decrease of the temperature.

For a non-relativistic degenerate electron gas, the electronic contribution to the specific heat is

$$
C_v^{(\text{elect})} = \frac{3}{2} \frac{k}{AH} \cdot \frac{\pi^2}{3} Z \frac{kT}{\varepsilon_F}
\tag{3}
$$

where Z, A are respectively, the atomic charge and mass, $k = 1.38 \times 10^{-16}$ erg $(K)^{-1}$ is Boltzmann's constant, and $H = 1.66044 \times 10^{-24}$g is the unit of atomic mass. Because of the high degeneracy kT is much less than the Fermi energy

$$
\varepsilon_F = \frac{(3\pi^2)^{2/3}}{2} \frac{\hbar^2}{m_e} \left(\frac{\varrho}{\mu_e H} \right)^{2/3}
$$

of the electrons, and $C_v^{(\text{elect})}$ is ignored in comparison with the specific heat of the ions (Approximation 3), which for a non-interacting ion gas becomes (Approximation 4)

$$
C_v^{(\text{ion})} = \frac{3}{2} \frac{k}{AH}.
\tag{4}
$$

The high degeneracy in the core of a white dwarf also promotes highly efficient heat conduction by the degenerate electrons, as was first shown by Marshak (1940). The core is therefore very nearly isothermal (Approximation 5), so that one finally obtains

$$
L \approx -\frac{3}{2} \frac{kM}{AH} \frac{\partial T_c}{\partial t},
\tag{5}
$$

where T_c is the core temperature.

In order to calculate T_c, one must deal with the problem of heat transfer through the thin, non-degenerate envelope of the white dwarf. If the envelope is in radiative equilibrium (Approximation 6), and if Kramers' law, $K = K_0 \varrho T^{-3,5}$ is used to represent the opacity (Approximation 7), the envelope equations can be integrated analytically, giving for the 'radiative, zero' surface conditions, $P = 0$ at $T = 0$, the result (Schwarzschild, 1958, p. 91)

$$\frac{1}{8.5} T^{8.5} = \frac{3}{4ac} K_0 \frac{\mu H}{k} \frac{L}{4\pi GM} \frac{1}{2} P^2, \tag{6}$$

where μ is the mean molecular weight of the envelope and P is the pressure. Since at the boundary of the isothermal, degenerate core the pressure and temperature are related through the condition $kT \approx \varepsilon_F$ (or $\varrho/\mu_e \approx 2.4 \times 10^{-8} T^{3/2}$: Schwarzschild, 1958, p. 60), Equation (6) can be reduced to a relation between core temperature and luminosity:

$$\frac{L}{L_\odot} \approx 1.7 \times 10^{-3} \frac{M}{M_\odot} \left(\frac{4 \times 10^{23}}{K_0} \right) \frac{\mu}{\mu_e^2} T_{c,7}^{3,5}, \tag{7}$$

where $\mu_e = A/Z$ is the mean molecular weight per electron, and $T_{c,7}$ is the core temperature in units of 10^7 K. With the aid of the result (7), Equation (5) can be integrated directly to give the age-luminosity law

$$\tau \approx \frac{7.6 \times 10^7}{A} \left(\frac{K_0}{4 \times 10^{23}} \frac{\mu_e^2}{\mu} \right)^{2/7} \left(\frac{M}{M_\odot} \right)^{5/7} \left(\frac{L}{L_\odot} \right)^{-5/7} y. \tag{8}$$

Equations (7) and (8) contain the main results of Mestel's theory. The principal approximations of the theory are summarized in Table I, and we next turn to a detailed investigation of the validity of these assumptions.

TABLE I

Main approximations in Mestel's model

1. Neglect of nuclear energy sources and sinks.

2. $T\dfrac{\partial s}{\partial t} \approx C_v \dfrac{\partial T}{\partial t}$, $\dfrac{\partial \varrho}{\partial t} = 0$: Neglect of residual gravitational contraction.

3. $C_v \approx C_v^{(\mathrm{ion})}$: Neglect of electronic heat capacity.

4. $C_v^{(\mathrm{ion})} \approx \dfrac{3}{2} \dfrac{k}{AH}$: Use of perfect gas law for ions.

5. $T(r) \approx T(0) \equiv T_c$: Isothermal core approximation.

6. Assumption of radiative equilibrium in envelope.

7. Use of Kramers' law opacity in envelope.

3. Nuclear and Neutrino Processes

A. NUCLEAR ENERGY PRODUCTION

Under conditions typical of the interior of a white dwarf ($\varrho \sim 10^6$ g·cm^{-3}, $T \sim 10^7$ K) hydrogen-burning nuclear reactions generate energy at an enormous rate, and – if the hydrogen abundance is at all appreciable – generate far more energy than is required to maintain the luminosity of the star. The condition that the total rate of energy production not exceed the observed white dwarf luminosity is alone sufficient to show $X_H < 10^{-4}$ (Marshak, 1940). For normal white dwarfs, however, much more stringent limitations are set by the following stability considerations (Schatzman, 1958; Schwarzschild, 1958; Mestel, 1965):

Suppose first that nuclear energy production occurs near the base of the non-degenerate envelope. As shown by Ledoux and Sauvenier-Goffin (1950), this situation leads to pulsational instability. For most white dwarfs this possibility is therefore ruled out as a significant contributor to the luminosity by the negative results of the recent observational search for pulsations in degenerate stars by Ostriker and his colleagues (Lawrence *et al.*, 1967). Nuclear burning must thus take place, if at all, in the deep interior of a white dwarf, where it cannot drive pulsations. However, for white dwarfs with $M \gtrsim 0.1\ M_\odot$ (the approximate lower mass limit for the hydrogen-burning main sequence), all hydrogen in the deep interior *must* already have been consumed. Furthermore, if some residual hydrogen did remain, any appreciable nuclear burning in the degenerate core would result in a thermal runaway, since a rise in the temperature is not offset by a proportionate increase in the pressure in degenerate matter, and the resulting secularly unstable situation is not of interest to a study of stable white dwarfs.

It appears, then, that nuclear reactions are unimportant for 'normal' white dwarfs, and I shall not discuss them further.

B. NEUTRINO ENERGY LOSSES

Adams *et al.* (1963) and Inman and Ruderman (1964), have shown that decay of a photon (or 'transverse plasmon') into a neutrino-antineutrino pair becomes possible in the dense plasma of the interiors of stars in the later phases of stellar evolution where $\hbar\omega_p/kT \sim 1$. Here $\omega_p = [4\pi n_e e^2/m_e]^{1/2}$ is the electron plasma frequency. This condition is satisfied in the immediate pre-white-dwarf evolutionary stages. Vila (1965) has studied the effect of this energy loss mechanism upon the evolution of a $1\ M_\odot$ pure iron star. More recently similar calculations have been done for iron stars of other masses by Savedoff *et al.* (1969), for carbon stars by Beaudet and Salpeter (1969) and by Kutter and Savedoff (1969), and for stars composed of 80% O, 10% Ne, 10% Mg by Vila (1966, 1967).

The effect of the plasma neutrino energy loss upon the evolution of a star in the immediate pre-white-dwarf stages is to deplete the thermal energy store on a timescale $\tau_\nu \sim kT/AH\varepsilon_\nu$, where ε_ν is the energy loss rate in erg gm^{-1} sec^{-1}. At $T \sim 10^8$ K, $\varrho \sim 10^6$ g·cm^{-3}, the more recent, improved calculation of ε_ν by Beaudet *et al.* (1967) show $\tau_\nu \sim 3 \times 10^7\ A^{-1}$ yr. This may be compared with the rate of evolution in the absence of neutrino losses at a similar stage, as given by Equation (8): $\tau \sim 5 \times 10^7\ A^{-1}$ yr. The neutrino losses considerably accelerate the evolution in those phases where $\tau_\nu \leqslant \tau$. This is shown graphically for several of the pure iron star models of Savedoff *et al.* (1969) in Figure 1, where the ages of the stars are given as functions of their optical luminosities. At $10^{-2}\,L_\odot$ the neutrino models are younger by about a factor of 10, while even at $10^{-4}\,L_\odot$ their ages are still 30–40% less than the ages of the corresponding models computed without neutrino emission. (These calculations used the older neutrino rates of Inman and Ruderman (1964) that were too large by a factor of 4 (Zaidi, 1965).) The age differences computed with the corrected neutrino rates of Beaudet *et al.* (1967) would be about a factor of 2.5 at $10^{-2}\,L_\odot$ and about 10% at $10^{-4}\,L_\odot$.) Since the timescale of evolution is proportional to the total thermal

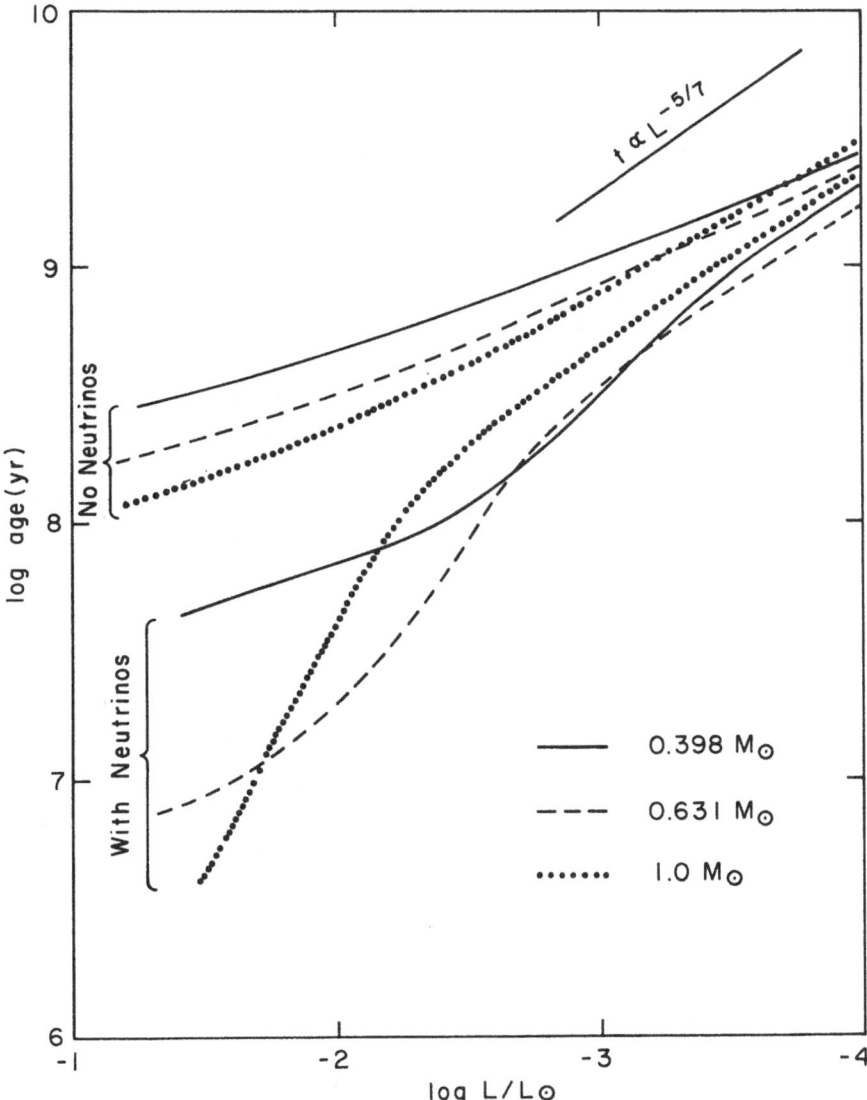

Fig. 1.   Effect of plasma neutrino emission on the ages of iron white dwarfs as computed by Savedoff *et al.* (1969). The slope of the age-luminosity relation $t \propto L^{-5/7}$ derived by Mestel (1952) is also shown.

energy content of 1 white dwarf (and thus to A^{-1}), no matter whether neutrino or photon emission is the dominant mechanism of energy loss, the ratio of ages of models with the same luminosity but with and without the inclusion of neutrino energy losses is independent of the chemical composition. Thus the effect is important for the brighter white dwarfs of all compositions if the direct electron-neutrino interaction actually does exist in nature, as predicted by Feynman and Gell-Mann (1958), and by Sudarshan and Marshak (1958), and accurate calculations of the ages of the white dwarfs must include the plasma mechanism of neutrino energy loss.

4. Gravitational Contraction and the Heat Capacity of the Electrons

The neglect of terms proportional to kT/ε_F in the equation of state of the degenerate electron gas is strictly justifiable only for rather high densities and low temperatures, conditions which are not met in the interiors of all white dwarfs; for example, at the average density of a 0.4 $M_\odot$ white dwarf $kT/\varepsilon_F > 0.1$ for $T > 2 \times 10^7$ K. The two main consequences of this are that for the less massive white dwarfs it is not sufficiently accurate to neglect either the residual gravitational contraction of the star or the contribution of the electrons to the specific heat of the stellar matter.

The order of magnitude of the contributions of these effects to net rates of energy release in the star can be established in the following way.

It follows from Equation (3) that the electronic contribution to the luminosity of a white dwarf is of the order of $\frac{1}{3}\pi^2 Z(kT/\varepsilon_F)$ times that of the ions. For reasonable chemical compositions the electrons can even contribute substantially more than the ions, in the hotter white dwarfs. This is shown in Figure 2 for the pure iron models of Savedoff et al. (1969), where the integrals over the entire star of $\varepsilon_{\text{ion}} = C_v^{(\text{ion})}\,\partial T/\partial t$, $\varepsilon_{\text{elect}} = C_v^{(\text{elect})}\,\partial T/\partial t$, and $\varepsilon_{\text{grav}} = -T\varrho^{-2}(\partial P/\partial T)_\varrho\,\partial\varrho/\partial t$ are plotted as functions of the stellar luminosity. It is evident that the approximation of neglect of the electronic heat capacity introduces errors of more than a factor of 2 in the lifetimes of the more luminous iron white dwarfs of all masses, while the effect continues to much lower luminosities for the stars of smaller mass. Because of the dependence of $C_v^{(\text{elect})}/C_v^{(\text{ion})}$ on Z, the use of an iron composition exaggerates this effect; however, even for a carbon composition, the electrons can still contribute as much as 30–50% of the luminosity of the hotter white dwarfs.

The contribution of the residual gravitational contraction to the energy balance is much smaller than the effect of the electronic heat capacity. We estimate the magnitude of this term using an argument due to Mestel and Ruderman (1967). They show that for white dwarfs of low mass

$$\int_0^M dM_r\, \frac{P^{(\text{elect})}}{\varrho^2}\, \Delta\varrho \approx -2\int_0^M dM_r \int_0^T C_v^{(\text{ion})}\, dT \equiv -2E_{\text{therm}}, \tag{9}$$

where $\Delta\varrho$ is the difference in density from the fully degenerate, 'black dwarf' state,

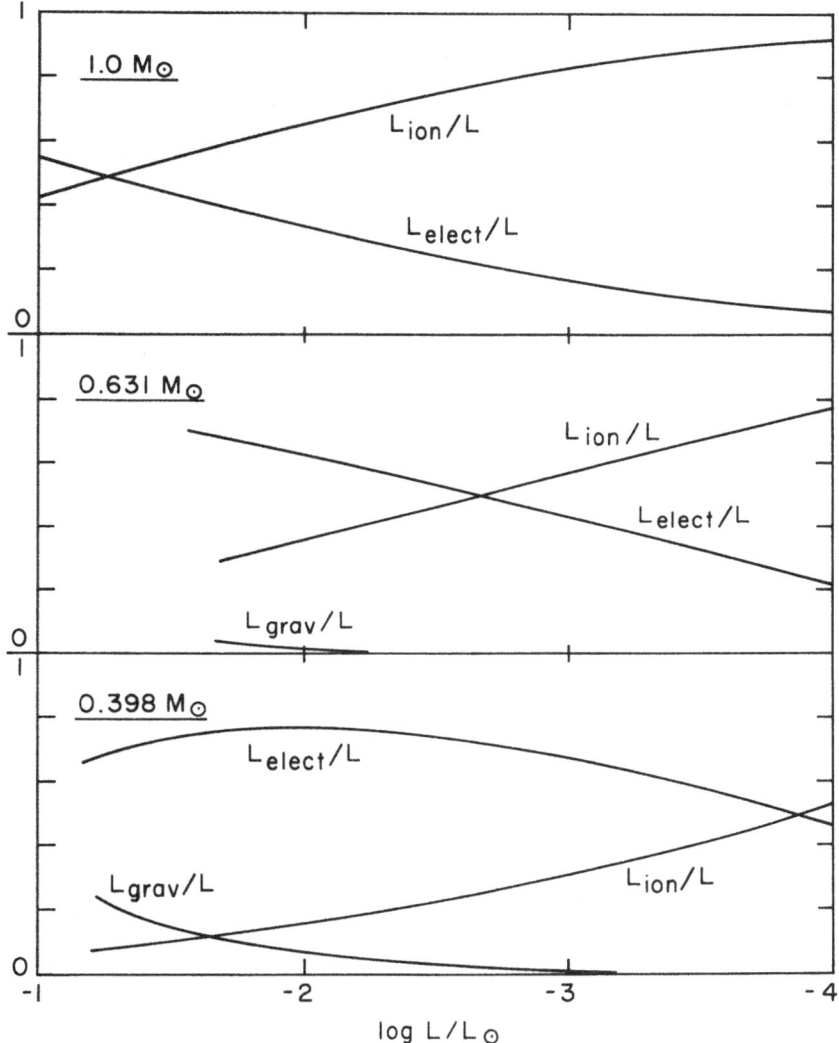

Fig. 2. Contributions of ion cooling, electron cooling and gravitational contraction to the luminosities of the iron white dwarfs without neutrino emission computed by Savedoff *et al.* (1969).

and we have neglected the coulomb interaction terms (which are discussed in the following section). As shown by Equation (2), the total 'gravitational' energy release is

$$\Delta E_{\mathrm{grav}} = - \int_0^M dM_r \frac{T}{\varrho^2} \frac{\partial P}{\partial T}\bigg|_\varrho \Delta\varrho = - \int_0^M dM_r \frac{P^{\mathrm{(elect)}}}{\varrho^2} \Delta\varrho \left[\frac{T}{P^{\mathrm{(elect)}}} \frac{\partial P}{\partial T}\bigg|_\varrho \right]. \quad (10)$$

Since the thermal correction to the degenerate electron pressure is $O(kT/\varepsilon_F)^2$, it is

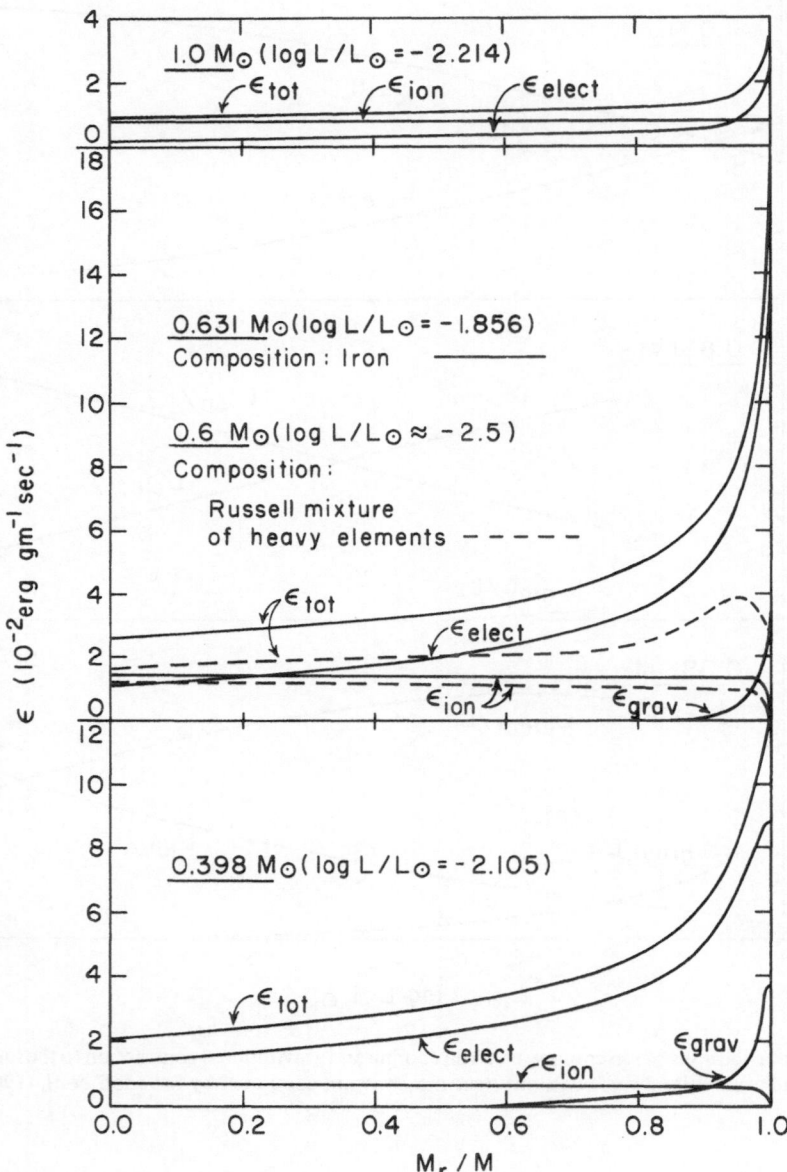

Fig. 3. Distribution of energy sources in white dwarfs: $\varepsilon_{\text{tot}} = -T\partial s/\partial t$, $\varepsilon_{\text{ion}} = -C_v^{(\text{ion})}\partial T/\partial t$, $\varepsilon_{\text{elect}} = -C_v^{(\text{elect})}\partial T/\partial t$, $\varepsilon_{\text{grav}} = -T\varrho^{-2}(\partial P/\partial T)_\varrho \, \partial\varrho/\partial t$. The solid curves refer to the iron white dwarf models without neutrino emission computed by Savedoff et al. (1969) and the dashed curves to the 0.6 $M_\odot$ white dwarf composed of the Russell mixture of heavy elements constructed by Hayashi et al. (1962).

negligible, and $\partial P/\partial T \mid_\varrho \approx k\varrho/AH$, from the ionic terms alone. Thus we have

$$\Delta E_{\text{grav}} \approx \left\langle \frac{P^{(\text{ion})}}{P^{(\text{elect})}} \right\rangle 2 \int_0^M dM_r \int_0^T C_v^{(\text{ion})} \, dT = \left\langle \frac{5}{Z} \frac{kT}{\varepsilon_F} \right\rangle E_{\text{therm}}, \tag{11}$$

where the average is defined by Equation (10).

The gravitational energy released thus can normally be neglected in a white dwarf, except in a star of very low mass (where $\langle kT/\varepsilon_F \rangle$ may be of the order of 10% or greater even at typical white dwarf luminosities), and in the outer, non-degenerate envelope. These conclusions are illustrated both by the ratio of the gravitational to total luminosity plotted in Figure 2, and by the distribution of energy sources in white dwarfs of different masses, as shown in Figure 3 for several of the models of Savedoff *et al.* (1969). Except in the low mass models, the gravitational term is completely negligible. Also shown in this figure is the distribution of energy sources in a $0.6\,M_\odot$ star composed of the Russell mixture of heavy elements, as computed by Hayashi *et al.* (1962, p. 164). The agreement between the two results is quite close when the contributions $\varepsilon_{\text{elect}}$ are scaled by the ratio of average ionic charges as indicated by Equation (3).

5. Effects of Coulomb Interactions

The recent advances in our understanding of the thermodynamic state of the interior of a white dwarf stem from the recognition – independently by Kirzhnits (1960), Abrikosov (1961), and Salpeter (1961) – that coulomb interactions at the low temperatures characteristic of the white dwarfs force the ions in the plasma to form a crystalline solid. The thermodynamic properties of the stellar plasma in these stages of evolution are dependent upon two dimensionless parameters, the ratio Γ of the typical coulomb interaction energy to kT and the ratio of the characteristic energy of lattice vibrations ('phonons') to kT. For a plasma containing only one species of ion, these quantities become

$$\Gamma \equiv \frac{(Ze)^2}{akT} = 2.28 \frac{Z^2}{A^{1/3}} \frac{\varrho_6^{1/3}}{T_7} \tag{12}$$

and

$$\frac{\hbar\Omega_P}{kT} = 2.240 \frac{\Theta}{T} = 2.240 \times 0.174 \left(\frac{2Z}{A}\right) \frac{\varrho_6^{1/2}}{T_7}, \tag{13}$$

where $\varrho_6 = \varrho/10^6 \text{ g}\cdot\text{cm}^{-3}$, $\frac{4}{3}\pi a^3 = (\varrho/AH)$ defines the radius a of a sphere which on the average contains one single ion, $\Omega_p^2 = 4\pi(\varrho/AH)(Ze^2)/AH$ is the square of the ion plasma frequency, and Θ is the Debye temperature of the solid.

These two parameters separate the evolution of a cooling white dwarf naturally into the following four stages*: (i) $\Gamma < \Gamma_m$, $T > \Theta$; in this stage the ions of the plasma

---

* Mestel and Ruderman (1967), applying the Lindemann melting point rule, found the transition between the high temperature coulomb liquid and the low temperature coulomb solid to occur at $\Gamma = \Gamma_m \approx 64$. A somewhat more accurate treatment of this calculation gave $\Gamma_m \approx 170$ (Van Horn, 1969). The numerical 'experiments' of Brush *et al.* (1966) yielded $\Gamma_m \approx 125$.

form a coulomb liquid. (ii) $\Gamma \gtrsim \Gamma_m$, $T > \Theta$; the ions crystallize into a regular lattice structure. (iii) $\Gamma > \Gamma_m$, $T < \Theta$; the specific heat of the lattice decreases rapidly to zero. (iv) $\Gamma < \Gamma_m$, $T < \Theta$; in this case the zero point motion of the ions is sufficient to prevent crystallization, and the final, zero-temperature state of the plasma is that of a charged quantum liquid instead of a solid.

I shall briefly describe the features of each of these regimes and discuss the attendant changes in the rates of evolution of the white dwarfs.

A. COULOMB LIQUID STAGE: $\Gamma < \Gamma_m$, $T > \Theta$

In Figure 4 the locus of points in the H−R diagram where $\Gamma = 100$ are shown for white dwarfs composed entirely of ^{4}He, ^{16}O, or ^{56}Fe, together with the observational data. It is apparent that Coulomb effects become extremely important in just the region occupied by the white dwarfs. Also shown in this figure are the 'experimental' data of Brush et al. (1966) for the coulomb corrections to the pressure and entropy of an ideal gas of ions, together with semi-empirical fits to the data which are given by

$$\frac{(P - P_0) V}{NkT} = - 0.113 \, \Gamma^{3/2} \left[\frac{1}{(1 + 0.142 \, \Gamma)^{1/2}} + \frac{1.54}{(1 + 0.575 \, \Gamma)^{3/2}} \right], \quad (14)$$

$$\frac{S - S_0}{Nk} = - \ln \left[1 + \frac{\Gamma^{3/2}}{2\sqrt{3}} \left(0.015 + \frac{0.585}{1 + \Gamma^{1/2}} + \frac{0.400}{1 + 1.308 \, \Gamma^{3/2}} \right) \right], \quad (15)$$

where P_0, S_0 are the pressure and entropy of a perfect gas of N ions in a box of volume V at temperature T. These expressions reduce to the correct, analytical results for $\Gamma \ll 1$ and $\Gamma \gg 1$ and are accurate to better than 10% over the range where the corrections are appreciable.

The coulomb correction terms have two effects upon the lifetimes of the white dwarfs. First, the correction to the entropy leads to a gradual increase in the specific heat from $\frac{3}{2} k/AH$ to $3 k/AH$ as the temperature drops toward the freezing point of the plasma. For the large Γ values typical of white dwarfs the heat capacity is very nearly $3 k/AH$ over the whole range of interest, and as first pointed out by Mestel and Ruderman (1967) this leads to a factor of two increase in the white dwarf lifetimes during this stage.

The second effect is to increase the rate of gravitational contraction above that indicated by Equation (11). With the aid of Equation (14) we find the contraction rate in the limit of large Γ to be about 1.7 times as great as given by Equation (11). For the low mass white dwarfs this effect is therefore not negligible.

B. STAGE OF CRYSTALLIZATION: $\Gamma \gtrsim \Gamma_m$, $T > \Theta$

When the temperature of the white dwarf core falls to the point where $\Gamma = \Gamma_m$ at the center, crystallization of the plasma begins. As shown by Brush et al. (1966), a heat of fusion $T \Delta s \sim \frac{3}{4} kT/AH$ is liberated in the transition from the liquid to the solid phase. This release of energy increases the time spent in the corresponding region of the H−R diagram by as much as 50% above the time computed by Mestel and

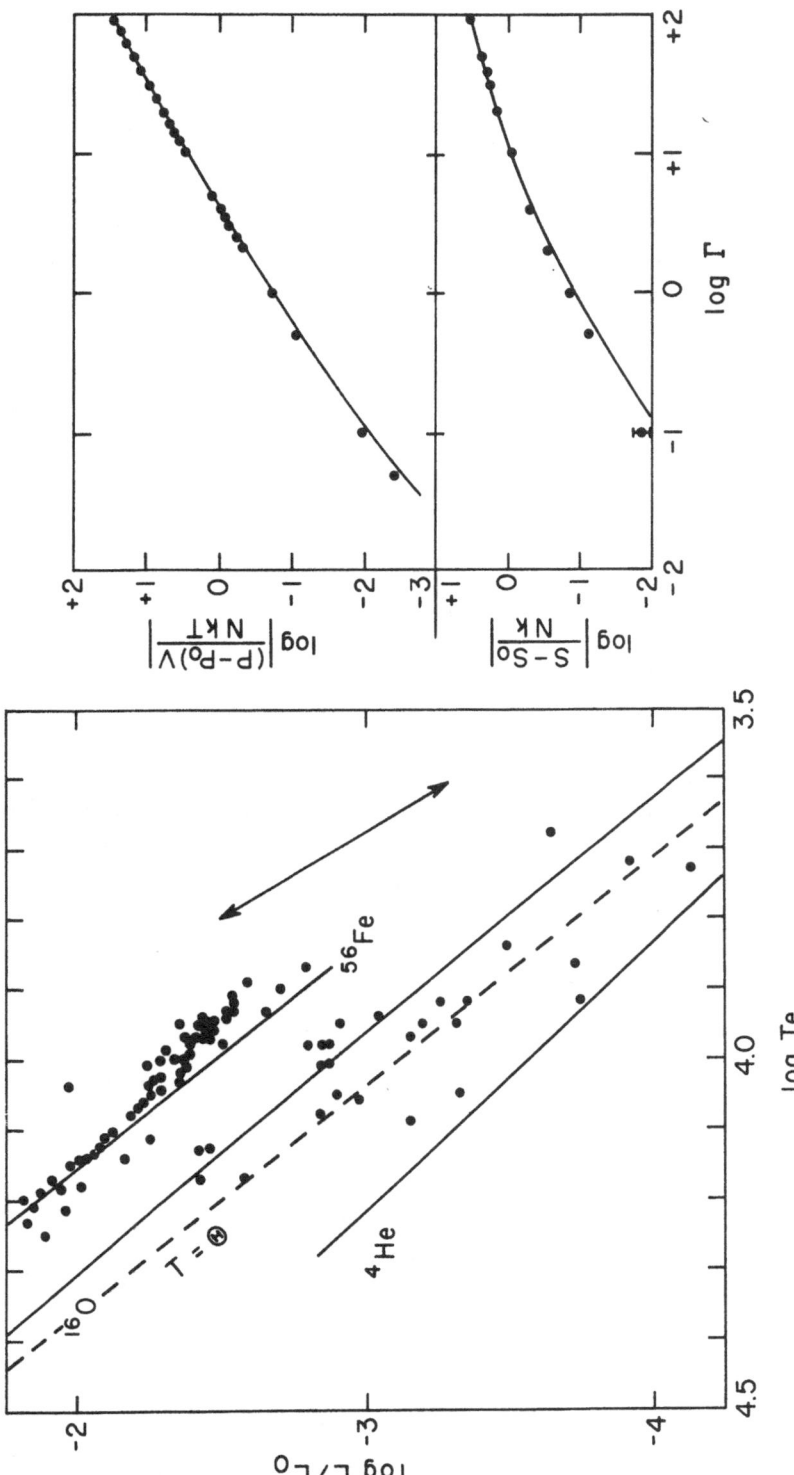

Fig. 4. Importance of coulomb corrections to the equation of state. The solid curves in the H − R diagram indicate the loci of points where $\Gamma = 100$ at the centers of ^{56}Fe, ^{16}O, and ^{4}He white dwarfs. The double-ended arrow shows the magnitude and direction of the shift in these curves produced by a factor of 2 change in Γ. The dashed curve indicates the locus of points where $T = \Theta$ at the center of the star; this curve is independent of the composition. Observed locations of the white dwarfs taken from Eggen and Greenstein (1965) are marked by solid dots. The magnitudes of the coulomb corrections to the pressure and entropy as calculated by Brush et al. (1966) are shown by the solid dots in the right-hand figure. Also shown are the semi-empirical analytic fits given by equations (14) and (15).

Ruderman (Van Horn, 1968). Because of the strong composition dependence of the Coulomb interaction, as shown, for example, by Equation (12), white dwarfs of different compositions undergo crystallization along different, rather well-defined sequences in the H−R diagram. It now appears doubtful, however, whether these sequences can be distinguished observationally (Weideman, 1969).

C. ONSET OF RAPID COOLING: $\Gamma > \Gamma_m$, $T < \Theta$

When T drops below the Debye temperature Θ of the lattice, excitation of the higher phonon energy levels becomes impossible, and the specific heat begins to fall, decreasing as

$$C_v^{(\text{ion})} \sim \tfrac{1}{5} 12\pi^4 (T/\Theta)^3 \, k/AH \quad \text{for} \quad T \ll \Theta$$

(Landau and Lifshitz, 1958, p. 187ff). The total thermal energy content given by Equation (9) is thus

$$E_{\text{therm}} = \int_0^M dM_r \int_0^T C_v^{(\text{ion})} \, dT \rightarrow \begin{cases} \dfrac{3k}{AH} T_c M, \, T_c \gg \Theta \\[2mm] \dfrac{3k}{AH} T_c M \cdot \dfrac{\pi^4}{5} \int_0^M \dfrac{dM_r}{M} \left(\dfrac{T_c}{\Theta}\right)^3, \, T_c \ll \Theta \end{cases} \tag{16}$$

The density dependence of Θ (Equation (13)) requires the integration over the entire mass to obtain the heat capacity of the entire star. The results of integrating the full Debye expression for the specific heat over the mass distributions for a number of white dwarfs have been tabulated by Van Horn (1968), and the substantial reduction in the lifetimes of the fainter white dwarfs has been discussed by Mestel and Ruderman (1967), by Van Horn (1968), and by Ostriker and Axel (1969). The latter paper shows that the inclusion of this effect leads to ages $< 10^{10}$ yr for all white dwarfs more massive than about 0.7 $M_\odot$.

The upper boundary of the region of rapid cooling in the H−R diagram is indicated in Figure 4, where the locus of points for which $T = \Theta$ at the center of the star is shown.

D. QUANTUM LIQUID STAGE: $\Gamma < \Gamma_m$, $T < \Theta$

When the temperature falls below Θ while $\Gamma < \Gamma_m$, the zero-point motions of the ions are sufficient to prevent the plasma from crystallizing, and the final, zero-temperature state is then that of a quantum liquid. The exact value of the stellar mass at which this transition occurs depends upon the value of Γ_m and is smaller for larger Γ_m. Because of the rather limited range of mass and composition for which white dwarfs can possess quantum liquid cores this stage of evolution has been largely neglected. However, see Abrikosov (1961).

6. Radiative Transfer near the Core-Envelope Boundary

The thermal structure of a typical white dwarf is shown in Figure 5, where the model of Sirius B computed by Marshak (1940) is given. At the boundary of the degenerate core ($\psi=0$), which is reached at a depth of one percent of the stellar radius, the temperature has already climbed to about half the value at the center of the star. It is at approximately this point also that the change from radiative transport to electron conduction occurs. This is indicated by the run of radiative and conductive opacities also shown in the figure. Since about half of the total temperature rise takes place in the non-degenerate envelope while most of the remaining change occurs in the outer part of the degenerate core, it is important to have accurate knowledge of both the

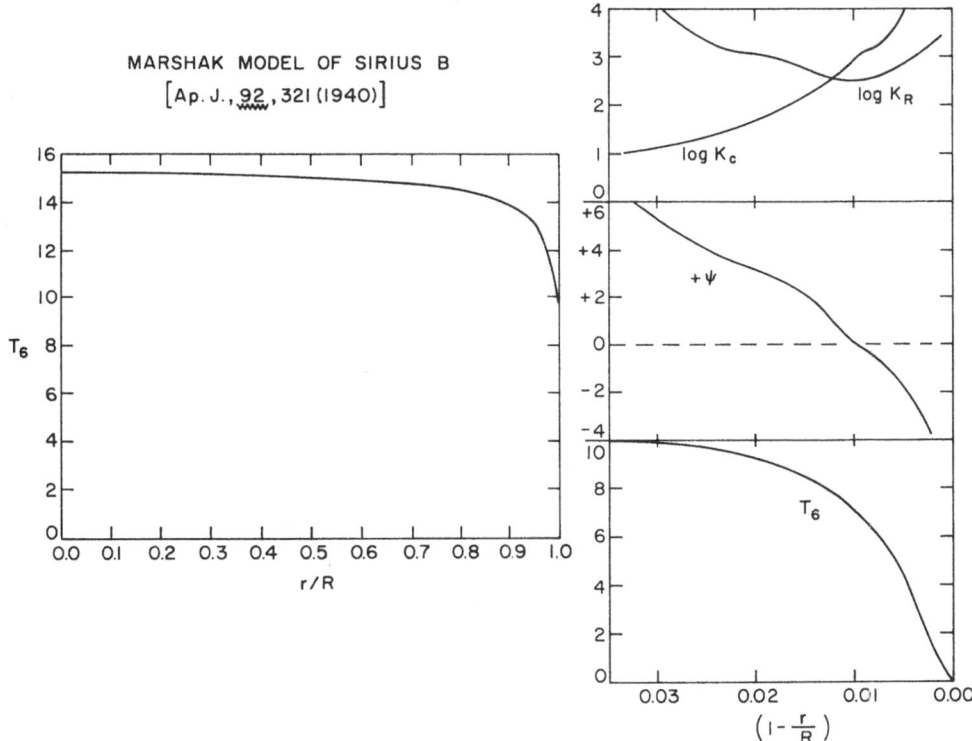

Fig. 5. Temperature distribution in Sirius B, as computed by Marshak (1940): $T_6 = T/10^6$ K, and R is the stellar radius. In the right-hand figure are plotted the distributions of temperature, of the degeneracy parameter ψ (ψkT is the Gibbs free energy per electron), and of the radiative, K_R, and conductive, K_C, opacities near the boundary of the degenerate core.

radiative and conductive opacities in order to carry out a precise computation of the temperature distribution and thermal energy content of a white dwarf. The assumption of a strictly isothermal core interior to the point where $\psi=0$ clearly leads to an underestimate of the total thermal energy and thus of the age by a factor of order two.

A. CONDUCTIVE OPACITIES

The high efficiency of heat conduction by the degenerate electrons in the interior of a white dwarf was established by Marshak (1940) and extended to the case of partial degeneracy by Mestel (1950) and Lee (1950). In this group of theories, conventionally termed the MML theory of the conductive opacity, the electrons are assumed to be non-relativistic and to be scattered only by the ions, which are taken to be infinitely massive point charges distributed at random through the plasma. While this process does in fact, provide the main contribution to the thermal resistivity, a number of other processes may also be significant in certain cases, as discussed by Abrikosov (1964) and by Hubbard (1966).

A substantial improvement in the treatment of electron conduction was made by Hubbard (1966), who carried out a rigorous calculation of the electron-ion scattering probability for a plasma and provided an approximate justification of the *ad hoc* cutoff procedure involved in the MML theory to avoid the divergence of the scattering cross-section associated with the long range of the coulomb force. Hubbard pointed out that the correlations among the positions of the ions induced by the Coulomb forces screens out the long-range electron-ion interactions, thus yielding a naturally convergent result. For typical white dwarf conditions Hubbard's opacities tend generally to be somewhat lower than corresponding MML opacities; the differences are of the order of a factor of two.

The effect of electron-electron scattering upon the conductivity has been studied intensively by Lampe (1968) and by Hubbard and Lampe (1969). These processes are normally neglected because the Pauli exclusion principle restricts the phase space available to *both* scattered electrons and thus inhibits these processes relative to electron-ion scattering. However, as Lampe (1968) has shown, near the core-envelope boundary, where the degeneracy is not strong, the electron-electron term can reduce the conductivity by 25–50%. When this process is taken into account, as in the extensive tabulations of Hubbard and Lampe (1969), the resulting opacity values are found to agree extremely well with the MML results in the degenerate core (the *opacity* is *in*creased above the value given by Hubbard), but is about a factor of two larger than the MML result in the non-degenerate region.

An important and badly needed extension of the Hubbard-Lampe theory to the conductivity of *relativistic*, semi-degenerate electrons has been published recently by Canuto (1970). At a density of 10^6 g·cm^{-3}, approximately the crossover from non-relativistic to relativistic degeneracy, the agreement with the Hubbard-Lampe results is quite close.

Two other processes that have also been considered are scattering from impurities (Hubbard and Lampe, 1969) and from phonons (Solinger, 1970) both of which can take place in the solid phase. Neither of these processes contribute significantly to the conductivity in a white dwarf, however.

B. RADIATIVE OPACITIES

The most extensive tabulations of radiative opacity data for astrophysically interesting element mixtures are those published by Cox and Stewart (1965). These calculations include contributions from large numbers of absorption lines as well as from bound-free, free-free, and electron scattering transitions. Corrections for partial electron degeneracy and plasma screening of the ionic potentials are included. Recently, Carson *et al.* (1968) have also attacked the problem of opacity calculations, but by a rather different method, using the high temperature Thomas-Fermi-Dirac atom model. For the two element-mixtures they have treated in common with Cox and his collaborators they find close agreement in the limit of high ionization. For low ionization,

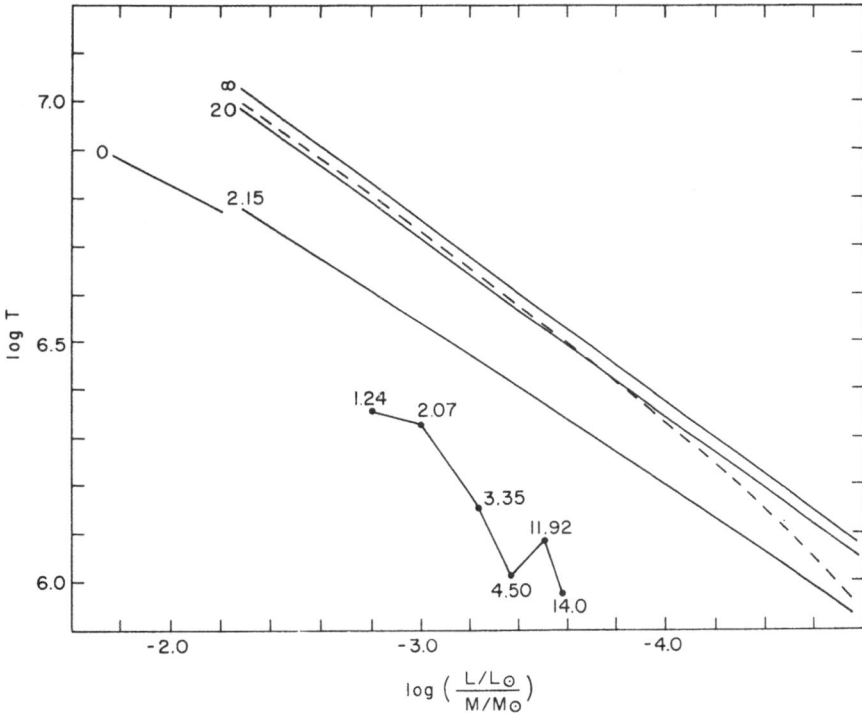

Fig. 6. Effects of different opacity laws and of envelope convection upon the core temperatures of white dwarfs. The solid curves show for helium envelopes the variation with L/M of the temperature at different values of ψ (the numbers attached to the curves indicate the appropriate ψ values), as computed by Lacis (1970) for MML conductive opacities and Kramers' radiative opacities. The dashed curve, also due to Lacis, is the temperature at $\psi = 20$ calculated with Hubbard-Lampe conductive opacities and Cox-Stewart radiative opacities. The temperature distribution at $\psi = 0$, is for a set of helium envelope models that include convection, as computed by Van Horn (1970). The jointed curve gives the distribution of temperatures at the base of the convective zone, as computed by Böhm (1970); the numbers along the curve refer to the ψ values at the bottom of the convective zone.

however, their opacities are larger by about a factor of two. This may be significant when envelope convection becomes important and the origin of the differences between these results should be investigated carefully.

Two additional effects that may become important in white dwarf envelopes are (i) the effect of plasma dispersion upon the Rosseland mean (Cox and Giuli, 1968) and (ii) the effect of the coulomb-induced ion correlations upon the processes of radiation absorption (Watson, 1970). Current estimates put both of these effects at about the ten percent level in typical white dwarf envelopes, although in certain cases they may be somewhat larger. More work on the difficult problem of opacity computations is thus desirable, but for purposes of establishing the relation between core temperature and luminosity the present calculations are probably adequate, at least for the white dwarfs of higher luminosity.

A detailed investigation of the effects on the L-T_c relation of differences in the opacity laws has recently been carried out by Lacis (1970). The results of his study are briefly summarized in Figure 6, where the temperatures at degeneracies of $\psi = 2.15$, 20, and ∞ are plotted as functions of the luminosity-to-mass ratio that characterizes the envelopes. These calculations made use of Marshak's (1940) version of Kramers' radiative opacity law and of the MML conductive opacity. Also shown in this figure are the temperatures at $\psi = 20$ in a model envelope computed using the Cox and Stewart (1965) radiative opacities for an almost pure helium mixture and the Hubbard and Lampe (1969) conductive opacities. The results agree to within about ten percent, indicating that present knowledge of the opacity laws is probably sufficient to provide a satisfactory determination of the L-T_c relation. This conclusion differs from that expressed by Hubbard and Wagner (1970) for reasons outlined below.

7. Envelope Convection in White Dwarfs

Although it has been known for some time that the envelopes of white dwarfs should be convectively unstable (Schatzman, 1958, p. 47; Kolesov, 1964), it was not until recently that Böhm (1968, 1969) pointed out that the extensive convective zones of the cooler white dwarfs result in appreciably lower core temperatures than expected on the basis of the radiative envelope model. In Figure 6 we have plotted the temperatures, obtained for nearly pure helium envelopes by Böhm (1970), at the inner boundaries of the convective zones (labeled by the degeneracy parameters ψ) as a function of L/M. It is evident that the core temperatures are substantially reduced; the reduction is about a factor of 3 at the lowest luminosity considered. The jog in this curve is due to the fact that the models with lower luminosities have *two* convection zones, the outer zone corresponding to the region of hydrogen and first helium ionization, while the deeper zone coincides with that of second helium ionization.

The coincidence of convective regions with partial ionization zones is a general phenomenon arising from the combination of very high opacities and reduced adiabatic gradients caused by the incomplete ionization (Hayashi *et al.*, 1962, p. 80). Convection is thus expected in the outer envelopes of all white dwarfs. This is confirmed by preliminary calculations for pure hydrogen, helium, and carbon envelope models (Van Horn, 1970). It is found that subsurface convection, *beginning at the photosphere*, occurs in all white dwarfs, but that the convective region occupies a smaller fraction of the envelope for the more luminous white dwarfs, while it may extend into the degenerate core at the lower luminosities. In the latter cases, the core temperature becomes quite sensitive to the actual atmospheric boundary conditions used; the models of Van Horn (1970) give much lower core temperatures in this luminosity range than the considerably more accurate models of Böhm (1970) which make use of actual atmospheric models computed by Grenfell and Böhm (1970). To obtain a reliable estimate of the L-T_c relation at such low luminosities it is, therefore, very important to have accurate opacity laws and atmospheres.

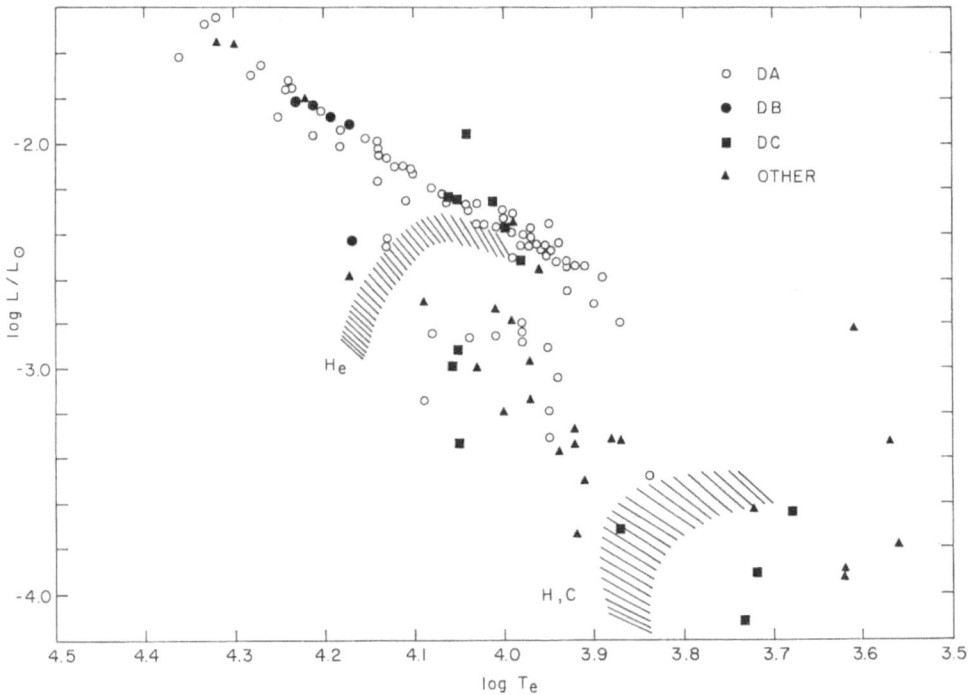

Fig. 7.   Boundaries in the H − R diagram of regions where envelope convection becomes important, shown together with the observations of Eggen and Greenstein (1965), with spectral types indicated. Below the upper hatched area marked 'He' the convective zone in a helium envelope extends into the degenerate core. Similar results hold for the hydrogen and carbon convection boundaries, indicated by the lower hatched area marked 'H, C'.

At the higher luminosities, where convection does not extend in to $\psi = 0$, the temperature distribution near the core-envelope boundary is independent of the surface conditions. This is a consequence of the well-known tendency of envelope models to converge to the so-called 'radiative zero' solution (Schwarzschild, 1958, p. 92). The effect is shown graphically by Lacis (1970) and by Van Horn (1970) and is indicated in Figure 6 by the distribution of temperatures at $\psi = 0$ obtained for helium envelopes with extensive convection zones in the outer parts. The good agreement of the temperatures of these models with results of the radiative envelope calculations thus indicates that prospects for obtaining accurate L-T_c relations for the brighter white dwarfs are much better than anticipated by Hubbard and Wagner.

Finally, it is of interest to ask in what regions of the H – R diagram we may expect envelope convection to be important. This is qualitatively answered in Figure 7, where the loci of points defined by the condition that the inner boundary of the convective zone just reach the edge of the degenerate core are plotted for envelopes composed of hydrogen, of helium, and of carbon. It is interesting to note that no white dwarfs of spectral type DB (strong helium lines) lie below the He-convection boundary, and that no DA stars (strong hydrogen lines) fall below the edge of the H, C-zone. A refined theory of white dwarf envelope convection such as that being developed by Böhm and his coworkers may soon be able to tell us whether this result is real and significant.

Acknowledgements

This work has been supported in part by the National Science Foundation under grant GP-13695. Some of the numerical calculations were carried out at the Computing Center of the University of Rochester, which is in part supported by National Science Foundation grant GJ-828.

References

Abrikosov, A.: 1961, *Soviet Phys.–JETP* **12**, 1254.
Abrikosov, A.: 1964, *Soviet Phys.–JETP* **18**, 1339.
Adams, J. B., Ruderman, M. A., and Woo, H. C.: 1963, *Phys. Rev.* **129**, 1383.
Beaudet, G., Petrosian, V., and Salpeter, E. E.: 1967, *Astrophys. J.* **150**, 979.
Beaudet, G. and Salpeter, E. E.: 1969, *Astrophys. J.* **155**, 203.
Böhm, K. H.: 1968, *Astrophys. Space Sci.* **2**, 375.
Böhm, K. H.: 1969, in *Low Luminosity Stars* (ed. by S. S. Kumar), Gordon and Breach, New York, p. 393.
Böhm, K. H.: 1970, *Astrophys. J.* **162**, 919.
Brush, S. G., Sahlin, H. L., and Teller, E.: 1966, *J. Chem. Phys.* **45**, 2102.
Canuoto, V.: 1970, *Astrophys. J.* **159**, 641.
Carson, T. R., Mayers, D. F., and Stibbs, D. W. N.: 1968, *Monthly Notices Roy. Astron. Soc.* **140**, 483.
Cox, A. N. and Stewart, J. N.: 1965, *Astrophys. J. Suppl.* **11**, 22.
Cox, J. P. and Giuli, T. T.: 1968, *Principles of Stellar Structure*, Gordon and Breach, New York, Chapter 8.
Eggen, O. J. and Greenstein, J. L.: 1965, *Astrophys. J.* **141**, 83.
Feynman, R. P. and Gell-Mann, M.: 1958, *Phys. Rev.* **109**, 193.
Grenfell, T. C. and Böhm, K. H.: 1970, *Astrophys. J.* **161**, 1183.
Hayashi, C., Hoshi, R., and Sugimoto, D.: 1962, *Suppl. Prog. Th. Phys.*, # 22.
Hubbard, W. B.: 1966, *Astrophys. J.* **146**, 858.
Hubbard, W. B. and Lampe, M.: 1969, *Astrophys. J. Suppl.* **18**, 297.
Hubbard, W. B. and Wagner, R. L.: 1970, *Astrophys. J.* **159**, 93.
Inman, C. L. and Ruderman, M. A.: 1964, *Astrophys. J.* **140**, 1025.
Kirzhnits, D. A.: 1960, *Soviet Phys.–JETP* **11**, 365.
Kolesov, A. K.: 1964, *Soviet Astron.–AJ* **8**, 387.
Kutter, G. S. and Savedoff, M. P.: 1969, *Astrophys. J.* **157**, 1021.
Lacis, A.: 1970, unpublished Ph.D. thesis, University of Iowa.
Lampe, M.: 1968, *Phys. Rev.* **170**, 306.
Landau, L. D. and Lifshitz, E. M.: 1958, *Statistical Physics*, Addison-Wesley Publ. Co., Reading, Mass., p. 187.
Lawrence, G. M., Ostriker, J. P., and Hesser, J. E.: 1967, *Astrophys. J.* **148**, L161.
Ledoux, P. J. and Sauvenier-Goffin, E.: 1950, *Astrophys. J.* **111**, 611.

Lee, T. D.: 1950, *Astrophys. J.* **111**, 625.

Marshak, R. E.: 1940, *Astrophys. J.* **92**, 321.

Mestel, L.: 1950, *Proc. Camb. Phil. Soc.* **46**, 331.

Mestel, L.: 1952, *Monthly Notices Roy. Astron. Soc.* **112**, 583.

Mestel, L.: 1965, in *Stars and Stellar Systems* **8** (ed. by L. H. Aller and D. B. McLaughlin), University of Chicago Press, Chicago, Chapter 5.

Mestel, L. and Ruderman, M. A.: 1967, *Monthly Notices Roy. Astron. Soc.* **136**, 27.

Ostriker, J. P. and Axel, L.: 1969, in *Low Luminosity Stars* (ed. by S. S. Kumar), Gordon and Breach, New York, p. 357.

Salpeter, E. E.: 1961, *Astrophys. J.* **134**, 669.

Savedoff, M. P., Van Horn, H. M., and Vila, S. C.: 1969, *Astrophys. J.* **155**, 221.

Schatzman, E.: 1958, *White Dwarfs*, North-Holland Publ. Co., Amsterdam.

Schwarzschild, M.: 1958, *Structure and Evolution of the Stars*, Princeton University Press, Princeton.

Solinger, A.: 1970, paper presented at 132nd meeting Am. Astr. Soc.

Sudarshan, E. C. G. and Marshak, R. E.: 1958, *Phys. Rev.* **109**, 1860.

Van Horn, H. M.: 1968, *Astrophys. J.* **151**, 227.

Van Horn, H. M.: 1969, *Phys. Letters* **28A**, 706.

Van Horn, H. M.: 1970, *Astrophys. J.* **160**, L53.

Vila, S. C.: 1965, Ph.D. thesis, University of Rochester.

Vila, S. C.: 1966, *Astrophys. J.* **146**, 437.

Vila, S. C.: 1967, *Astrophys. J.* **149**, 613.

Watson, W. D.: 1970, *Astrophys. J.* **159**, 641.

Weidemann, V.: 1968, *Ann. Rev. Astron. Astrophys.* **6**, 351.

Zaidi, M. H.: 1965, *Nuovo Cimento* **40**, 502.

16. THE LINE SPECTRA OF WHITE DWARFS

P. A. STRITTMATTER and D. T. WICKRAMASINGHE

Institute of Theoretical Astronomy, Cambridge, U.K.

Abstract. Model atmospheres with temperatures in the range $10000 \leq T_{eff} \leq 25000$ K and with gravities $6 \leq \log(g) \leq 9$ have been constructed for various helium abundances with a view to understanding the spectra of the hotter white dwarfs. It is shown that the DB stars are confined to the temperature range $15000 \leq T_{eff} \leq 18000$ K in which convection also becomes important in the outer layers of helium rich stars. The DA stars seem to avoid this temperature range. The hydrogen and metal abundances in DB atmospheres are shown to be reduced by factors 10^5 and 10^3 respectively compared to solar values. Possible explanations of DB and DC spectra are discussed.

1. Introduction

A considerable amount of observational data on the photometric and spectroscopic properties of white dwarfs has now been assembled, largely by Eggen and Greenstein (1965, 1967), Eggen (1970), Greenstein (1960, 1969). On the other hand comparatively little progress has so far been made in the theoretical interpretation of this data, mainly perhaps due to the lack of an adequate description of atomic processes in high density plasmas. While certain theoretical difficulties still exist, particularly in computing opacities and ionisation equilibria at high densities, enough is now known to allow calculation of reasonably accurate model atmospheres for these stars. Such models have accordingly been computed by us for a range of effective temperatures, gravities and compositions. In this paper we shall use the computed line strengths and photometric data to analyse the gross features of the spectra of hot ($T_{eff} \gtrsim 10000$ K) white dwarfs. A detailed account of this work is given in Paper III (Strittmatter and Wickramasinghe, 1970).

2. The Model Atmosphere and Line Profile Calculations

For each set of input parameters (effective temperature, gravity and composition) a constant flux model atmosphere in radiative equilibrium has been computed using the basic procedure outlined by Avrett and Krook (1963). Where Lyman and Balmer line blanketing is important it has been taken into account using a modified temperature correction procedure (Wickramasinghe and Strittmatter, 1969). The effect of blanketing by He I lines on the structure of the atmosphere is negligible at the temperatures considered here because the peak continuum flux never coincides with the position of strong lines. In each case flux constancy was achieved to an accuracy of better than 1%. Some of the atmospheres turned out to be convectively unstable in certain optical depth ranges; the effect has not been taken into account in our present calculations since, to the accuracy required here, the resultant changes in the observable parameters have been shown to be small (Wickramasinghe and Strittmatter, 1970; Paper II).

Luyten (ed.), White Dwarfs, 116–123. *All Rights Reserved.*

The sources of continuum opacity included are H, H^-, H_2, He, He^+, He^- and electron scattering. The He^- absorption coefficient has been taken from Somerville (1965); the remainder are treated as specified by Vardya (1961). Absorption from the bound states of metals has not been taken into account although it could cause substantial blanketing at far ultra-violet wavelengths particularly in the hydrogen deficient stars. This omission will generally cause our models to appear slightly cooler than would otherwise be the case, but has been made (a) for computational economy and (b) because it appears from the observations that the metal abundance in the surface layers of most white dwarfs is very low.

The He I $\lambda4472$ line has been computed using the detailed profiles of Griem (1968). The forbidden component at $\lambda4517$ is included in the computation of the equivalent width and may result in over-estimation of this quantity as compared to the observations (Paper II). We have also computed the lines He I $\lambda4713$, He I $\lambda5875$, Mg II $\lambda4481$, Si II $\lambda4129$, Ca II (H + K), C II $\lambda4267$ and Fe I $\lambda4272$. The broadening mechanisms included in those calculations are van der Waal's broadening by He and H, quadratic Stark broadening, Doppler broadening and radiation damping; the latter two are generally negligible in white dwarf atmospheres.

The UBV photometric parameters have been computed from the theoretical energy distribution (continuum and hydrogen lines) using the prescription of Matthews and Sandage (1963).

The basic models may be divided into two groups. The first group consists of a

TABLE I

Photometry and equivalent widths for several models

$T_{eff}/10^3$	$\log(g)$	N_{He}	B − V	U − B	B.C.	T_{eff} W(H$_\gamma$)	W(He I $\lambda4472$)	V
10	8	0.144	0.21	−0.66	−0.56	30.2	0	12.86
12	8	0.144	0.19	−0.62	−0.70	49.9	0	12.21
15	8	0.144	0.07	−0.74	−1.12	48.5	0	11.66
20	8	0.144	−0.07	−0.94	−1.88	36.0	2.19	11.17
25	8	0.144	−0.17	−1.04	−2.47	28.0	6.27	10.79
10	8	100.0	0.06	−0.88	−0.79	11.6	0	13.09
15	8	100.0	0.02	−0.90	−1.25	36.4	4.94	11.79
20	8	100.0	−0.16	−1.04	−1.82	17.9	22.4	11.11
25	8	100.0	−0.27	−1.10	−2.40	0.1	22.3	10.72
10	8	1 000.0	0.01	−0.92	−0.88	3.45	0.0	13.17
15	8	1 000.0	−0.10	−1.04	−1.43	11.5	5.18	11.97
20	8	1 000.0	−0.23	−1.04	−1.79	1.58	29.4	11.08
25	8	1 000.0	−0.27	−1.08	−2.51	0	24.4	10.83
10	8	10 000.0	0.01	−0.92	−0.87	0	0	13.17
15	8	10 000.0	−0.15	−1.07	−1.53	1.06	5.99	12.07
	7		−0.15	−1.05	−1.47	3.78	8.86	12.01
20	8	10 000.0	−0.23	−1.03	−1.88	0	26.2	11.17
	7		−0.23	−1.01	−1.88	0	23.8	11.18
25	8	10 000.0	−0.27	−1.07	−2.58	0	29.2	10.9
	7	10 000.0	−0.27	−1.06	−2.58	0	22.7	10.9

grid of twenty model atmospheres with $\log(g)=6$, 7, 8, 9 and $T_{\text{eff}}=10\,000$, 12\,000, 15\,000, 20\,000, 25\,000 K and are of 'normal' composition ($N_{\text{He}}=0.144$, $n_m/n_{\text{H}}=0.00165$). The second group of fifteen model atmospheres has input parameters $\log(g)=8$, $T_{\text{eff}}=10\,000$, 15\,000, 20\,000, 25\,000 K, $N_{\text{He}}=10^2$, 10^3, 10^4; and $\log(g)=7$, $T_{\text{eff}}=15\,000$, 20\,000, 25\,000, $N_{\text{He}}=10^4$. The metal abundance in the second, helium rich group was fixed by taking $N_m=n_m/n_{\text{He}}=10^2$. A second computation of certain line strengths was carried out for the second group using metal abundances reduced by a factor 10 but the atmospheric structure was assumed to be the same as before.

In Table I we list photometry and equivalent widths of $H\gamma$ and $He\,I\,\lambda4472$ for a range of models with increasing helium content. The effects of the $He\,I$ lines on the UBV colours has not been included in these calculations (see discussion below).

In Figure 1 we have plotted equivalent widths of $H\gamma$ against $U-V$ for a range of models with $\log g=8$ but different helium abundances. For the normal DA stars equivalent widths of up to 50 Å occur, falling however, very sharply once H^- opacity sets in at 12\,000 K.

The $He\,I\,\lambda4472$ equivalent width is plotted against $U-V$ in Figure 2. Values of $W(4472)$ of nearly 30 Å are reached for the high helium abundances. The sharp drop-off in equivalent width for $T \lesssim 20\,000$ K in these models is due to He^- opacity. The influence of the latter source is maintained until $T=25\,000$ K in the case when $\log g=8$ and $N_{\text{He}}=10^4$, thus accounting for the continuing rise with temperature of $W(4472)$. For $\log g=7$, however, the He^- ceases to be of importance for $T\sim20\,000$ K

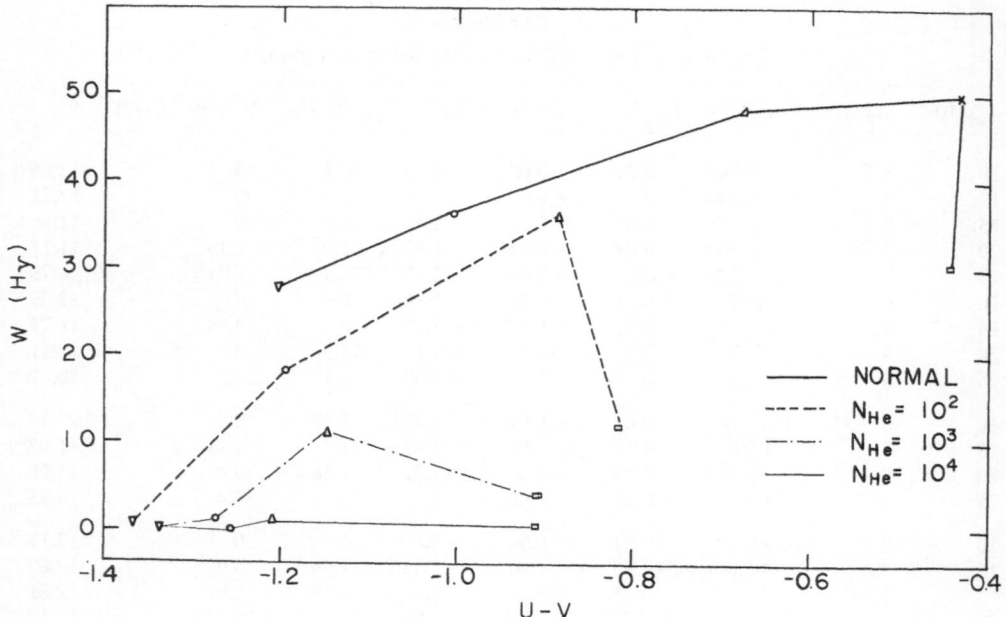

Fig. 1.   A plot of the equivalent width of $H\gamma$ against $(U-V)$ for the various models with $\log(g)=8$. The squares, triangles, circles and inverted triangles correspond to models with $T_e=10\,000$; 15\,000; 20\,000; 25\,000 K respectively. The symbol X is used to denote a $T_e=12\,000$ K model.

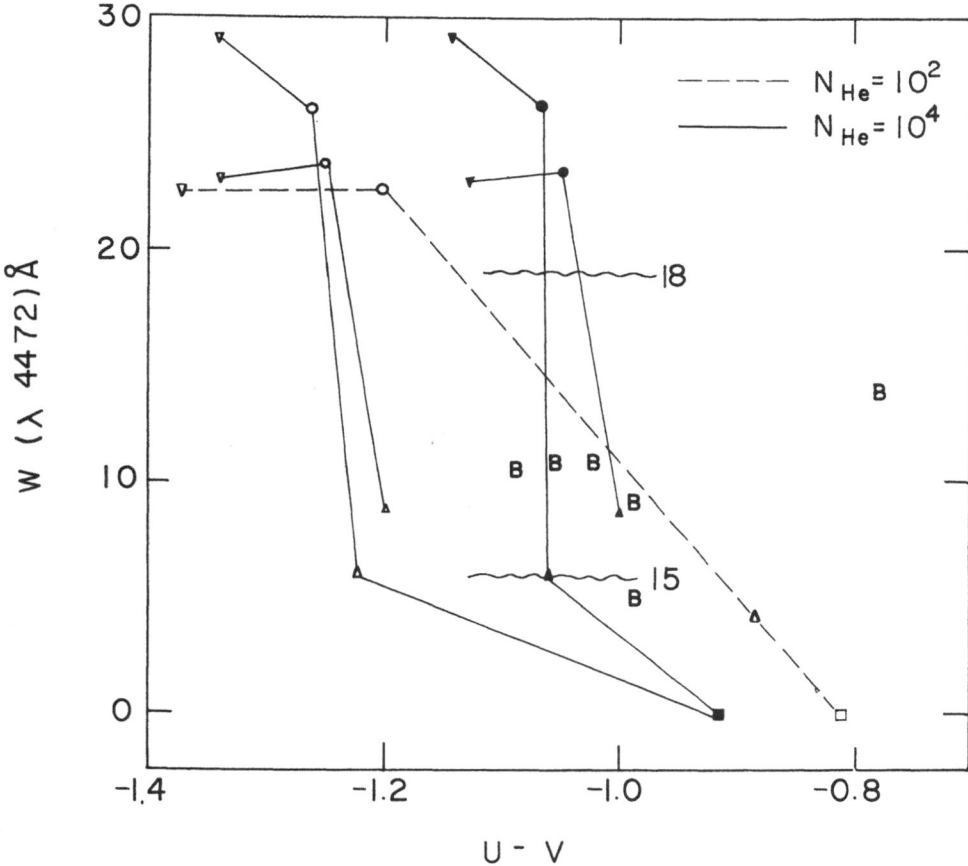

Fig. 2. A plot of the equivalent width of He I ($\lambda 4472$) against (U − V) for the helium rich models. Smaller symbols denote models with $\log(g) = 7$. All other models have $\log(g) = 8$. Open symbols denote colours computed without inclusion of the He I lines, solid symbols those in which the lines have been taken into account by the approximate method (see text). The letter B is used to denote the positions of the observed DBs with known W($\lambda 4472$). Otherwise the notation is as in Figure 1.

and W (4472) accordingly decreases with further increases in temperature owing to the decrease in the Boltzmann ratio of level populations giving rise to the He I line and continuum opacity respectively. From Table I we note that for intermediate compositions ($1 \lesssim N_{\text{He}} \lesssim 10^3$) both the H and He I lines are very strong. We may conclude from the fact that no such stars have yet been observed that this range of composition is very rare. Only when $N_{\text{He}} \gtrsim 10^4$ are the hydrogen lines negligible over the entire temperature range. White dwarf atmospheres are either hydrogen rich or extremely hydrogen deficient.

At this point, however, we should consider the effect of He I lines on the computed colours. It is quite clear from Figure 2 that W (4472) would be most easily observable for $N_{\text{He}} = 10^4$ at colours much bluer (U−V = −1.2) than those observed

$(U-V=-1.0)$. However, both the U and B bands contain large numbers of lines and these will clearly have the effect of reddening $U-V$ since only two HeI lines, $\lambda 5875$ and $\lambda 6678$ appear in the V band. The task, however, of computing each line individually in order to calculate broad band photometric parameters is not only immense but extremely misguided. Alternative observational parameters will therefore be given in Paper IV. Since much of our *present* knowledge is in the form of UBV data, however, we shall give an approximate method for *estimating* the blanketing effects.

Klemola (1961) has given line strengths of more than fifty HeI lines in the spectrum of the helium star $BD+10°\,2179$. We have accordingly taken these values, bunched them in 50 Å intervals and scaled them according to the ratio of our computed value of $W(4472)$ to that observed in $BD+10°\,2179$. These 'pseudo lines' were then included in the fluxes from which UBV photometry was computed. The fact that for more closely spaced lines the full effect of growth in equivalent width will not be obtained was taken into account in an approximate way. The resultant corrections to the colours of the various models are shown in Table II. We do not claim that they are

TABLE II

Corrections to colours for various models

T_{eff}	15000		20000		25000	
log g	7	8	7	8	7	8
$(U-B)$	-1.05	-1.07	-1.02	-1.03	-1.06	-1.07
$(B-V)$	-0.15	-0.15	-0.23	-0.23	-0.27	-0.27
$\Delta(U-B)$	$+0.11$	$+0.10$	$+0.06$	$+0.06$	0.06	0.05
$\Delta(B-V)$	$+0.09$	$+0.06$	$+0.14$	$+0.14$	0.14	0.15
$(U-B)_c$	-0.94	-0.97	-0.96	-0.97	-1.00	-1.02
$(B-V)_c$	-0.06	-0.09	-0.09	-0.09	-0.13	-1.12
$(U-V)_c$	-1.00	-1.06	-1.05	-1.06	-1.13	-1.14

accurate, merely that they give a reasonable estimate for the effect. While the corrections to $U-V$ and $B-V$ are substantial at *high* values of $W(4472)$, the change in $U-B$ is small. This results from the higher density of lines in the U band which accordingly reaches saturation at somewhat lower values of $W(4472)$ than does the B band. Thus while the initial $U-B$ change is strongly redwards it then reverses with increasing $W(4472)$ until both U and B bands are effectively 'saturated'. We note that the computed corrections cause a substantial reddening in $U-V$ for $T\gtrsim 15000$ K causing the drop in $W(4472)$ with colour to be even steeper than shown in Figure 2. It is also clear from Figure 2 that both the colours and equivalent widths of HeI $\lambda 4472$ agree well with the observations for $15000 \lesssim T \lesssim 18000$ K.

Finally we list in Table III the strengths of the following additional lines; HeI $\lambda 4713$, HeI $\lambda 5875$, MgII $\lambda 4481$, SiII $\lambda 4129$, CaII $(H+K)$, CII $\lambda 4267$ and FeI $\lambda 4272$. The

TABLE III

Strengths of additional lines

$\frac{T_{\rm eff}}{10^3}$	$\log g$	$N_{\rm He}$	HeI $\lambda 5875$	HeI $\lambda 4713$	MgII $\lambda 4481$	SiII $\lambda 4129$	CaII (H + K)	CII $\lambda 4267$	FeI $\lambda 4272$
			Normal metal abundance						
10	8	0.144	0	0	0.449	0.099	2.35	0	0.210
15	8	0.144	0.326	0.028	0.734	0.314	0.138	0.012	0
20	8	0.144	1.77	0.381	0.5	0.391	0.046	0.106	0
25	8	0.144	4.433	1.21	0.3	0.274	0.016	0.336	0
			High metal abundance ($N_m = 10^{-2}$)						
10	8	10000.0	0.028	0	5.63	1.6	80.4	0	16.3
15	8	10000.0	3.76	0.784	16.5	5.90	29.9	0.255	2.53
	7		4.29	1.68	15.64	5.90	16.5	0.462	0.783
20	8	10000.0	22.6	8.27	8.34	4.87	0.832	1.45	0.025
	7		14.8	5.68	3.82	2.88	0.249	1.04	0.008
25	8	10000.0	23.5	8.69	3.10	2.41	0.145	1.44	0
	7		13.5	5.03	1.22	1.16	0.070	0.988	0
			Low metal abundance ($N_m = 10^{-3}$)						
10	8	10000.0	0.028	0	1.55	0.307	28.9	0	5.22
15	8	10000.0	3.76	0.784	5.15	1.91	9.77	0.032	0.942
	7		4.29	1.68	4.94	2.11	5.27	0.104	0.283
20	8	10000.0	22.6	8.27	2.58	1.72	0.277	0.371	0.004
	7		14.8	5.68	1.20	1.04	0.085	0.334	0.0
25	8	10000.0	23.5	8.69	0.905	0.769	0.046	0.374	0.0
	7		13.5	5.03	0.391	0.369	0.026	0.322	0.0

results are for $\log g = 8$ and $N_{\rm He} = 0.144$ and 10^4 respectively, it having been already demonstrated that intermediate cases are rare if they exist at all. For the high helium abundance, results for two values of $N_m = 10^{-2}$ and 10^{-3} and for $\log g = 7$ are given.

3. Conclusions

It has been noted by Greenstein (1969) that the very existence of DB stars is remarkable if accretion of interstellar matter can occur. For a typical DB white dwarf, moving through the interstellar medium with a velocity of 30 km/sec, and accreting matter at the Eddington rate, the time taken to accrete 'one' atmosphere is $\sim 10^3$ yr. If a tail shock can form and destroy momentum perpendicular to the direction of motion of the star, this accretion time scale is reduced to $\sim 10^{-1}$ yr (Bondi and Hoyle, 1944). As we have shown in the previous section, only a small amount of hydrogen is sufficient to produce detectable hydrogen lines. We have accordingly investigated various mechanisms for preventing accretion. The roles of convection, circulation and gravitational diffusion are also examined (Paper III). Our main conclusions may be summarised as follows.

(i) The hydrogen abundance in DB white dwarfs must be at least 10^5 times smaller relative to helium than in normal stars.

(ii) The helium abundance in DA white dwarfs cannot exceed the normal value.

(iii) No intermediate hydrogen to helium ratios have yet been observed.

(iv) Metal abundances in DB stars must be at least a factor 10^3 lower than the solar value.

(v) Most DB stars lie in the temperature range $15000 \lesssim T_{\text{eff}} \lesssim 18000$ K at which convection becomes important in the outer layers.

(vi) DA stars apparently avoid the temperature range occupied by the DB stars.

(vii) Accretion should be important except in high velocity white dwarfs with moderately strong magnetic fields. It should occur in the DB stars.

(viii) Gravitational diffusion of heavy elements could be of importance in the presence of a magnetic field. In general a field strong enough to prevent accretion will allow diffusion to occur.

We suggest that these findings may be understood (hypothesis A) if most white dwarfs have helium rich envelopes and fairly weak magnetic fields. They therefore accrete interstellar material which resides on the surface and allows them to cool as normal DA stars. When the helium recombines convection is set up which removes the accreted material and mixes it downward. This surface convection has been shown to occur in the temperature range $15000 \lesssim T_{\text{eff}} \lesssim 18000$ K. The difficulties with this hypothesis, in particular the problems of stability of the cooler atmospheres and dilution of accreted material, lead to some doubts despite the satisfactory observational picture.

As an alternative we propose that accretion can be prevented in DB stars perhaps due to rotation coupled to a moderately strong magnetic field (hypothesis B). The concentration of stars in the range $15000 \lesssim T_{\text{eff}} \lesssim 18000$ K could then be attributed to the establishment of a deep convective zone in the envelopes of these stars (van Horn, 1970). The apparent absence of DA stars in this temperature range would, however, have to be attributed to chance in a small sample.

We also propose that there is a class of 'peculiar' white dwarfs, with moderately high space motions and above average magnetic fields. Accretion is thus prevented and diffusion of heavy elements downward can occur. These stars accordingly cool like DBs until the temperature falls below ~ 12000 K when the He I lines disappear. Provided metal diffusion is efficient enough the star would then appear as a hot DC-He$^-$ providing the opacity source. As on the main sequence 'peculiar' spectra would thus be associated with the presence of a magnetic field. On hypothesis B all DB stars could give rise to DC spectra at lower temperatures.

It appears to us that hypothesis A is perhaps in more obvious agreement with the observational evidence, but that hypothesis B raises fewer difficulties of principle (particularly since it does not demand large numbers of hydrogen and metal deficient, low velocity white dwarfs).

References

Avrett, F. H. and Krook, M.: 1963, *Astrophys. J.* **137**, 874.
Bondi, H. and Hoyle, F.: 1944, *Monthly Notices Roy. Astron. Soc.* **104**, 273.
Brechot-Sahal, S.: 1968, *Z. Astrophys.* **69**, 74.
Eggen, O. J.: 1970, *Astrophys. J.* **159**, 945.
Eggen, O. J. and Greenstein, J. L.: 1965, *Astrophys. J.* **141**, 83.
Eggen, O. J. and Greenstein, J. L.: 1967, *Astrophys. J.* **150**, 927.
Greenstein, J. L.: 1960, *Stars and Stellar Systems* **6** (ed. by J. L. Greenstein), University of Chicago Press, Chicago, p. 676.
Greenstein, J. L.: 1969, *Astrophys. J.* **158**, 281.
Griem, H.: 1968, *Astrophys. J.* **154**, 1111.
Klemola, A. R.: 1961, *Astrophys. J.* **134**, 130.
Mathews, T. A. and Sandage, A.: 1963, *Astrophys. J.* **138**, 30.
Somerville, W. B.: 1965, *Astrophys. J.* **141**, 811.
Strittmatter, P. A. and Wickramasinghe, D. T.: 1970, *Monthly Notices Roy. Astron. Soc.*, in press (paper III).
Terashita, Y. and Matsushima, S.: 1969, *Astrophys. J.* **156**, 183.
van Horn, H. M.: 1970, *Astrophys. J. Letters* **160**, 53.
Vardya, M. S.: 1961, *Astrophys. J.* **133**, 107.
Wickramasinghe, D. T. and Strittmatter, P. A.: 1969, *Astron. Astrophys.* **2**, 242.
Wickramasinghe, D. T. and Strittmatter, P. A.: 1970, *Monthly Notices Roy. Astron. Soc.*, **150**, 435 (paper II).

17. ON THE ORIGIN OF WHITE DWARFS

B. PACZYŃSKI*

Institute of Theoretical Astronomy, University of Cambridge

Model evolutionary calculations have been made for population I stars ($X=0.7$, $Z=0.03$) with masses of 0.8, 1.5, 3, 5, 7, 10, and 15 $M_\odot$ (Paczyński, 1970). Neutrino losses were taken into account. All the models were evolved up to the red supergiant phase with the hydrogen and helium burning shell sources. In this phase of evolution the hydrogen rich envelopes of 0.8, 1.5, and 3 $M_\odot$ stars became dynamically unstable due to the large depth of the hydrogen and helium ionization zones. It was assumed that almost entire envelopes were lost as a result of that instability. The degenerate carbon-oxygen cores of 0.6, 0.8, and 1.2 $M_\odot$ were left with the hydrogen rich envelopes of a small mass. Almost all the hydrogen that was left was subsequently burnt in the shell source. Finally, the hydrogen and helium burning shell sources disappeared as the nuclear fuel was exhausted. The models were cooling down to the white dwarf phase and they had a small amount of hydrogen left close to their surfaces. It is estimated that population I stars with masses up to 3.5 $M_\odot$ may produce white dwarfs with masses up to 1.37 $M_\odot$. The hydrogen rich envelopes that were lost as a result of dynamical instability could form planetary nebulae with masses up to 2 $M_\odot$. Further details are published in the paper referred to above.

Reference

Paczyński, B.: 1970, *Acta Astron.* **20**, 47.

* On leave from the Institute of Astronomy, Polish Academy of Sciences.

18. MODEL ATMOSPHERES
FOR HYDROGEN-DEFICIENT WHITE DWARFS

I. BUES

Institut für theoretische Physik und Sternwarte der Universität Kiel, Germany

Abstract. The determination of atmospheric parameters for non-DA white dwarfs is investigated with the computed helium-rich model atmospheres by Bues (1970). Only poor predictions are possible from UBV colors alone for DB and DC stars. From uvby colors a determination of effective temperature is possible within 1000 K. Profiles of lines in different parts of the spectrum are necessary for better results.

A deficiency of metal abundances for the cooler non-DA stars is obtained.

For hydrogen-deficient white dwarfs the determination of atmospheric parameters is more uncertain than for DA-stars, since the different chemical composition is not specified in advance and the absorption coefficients of elements other than hydrogen are only approximately available.

As a first step in model computation for non-DA white dwarfs we therefore computed line-free flux-constant model atmospheres for 13 different helium-rich compositions (ratio H:He varying from $1:10$ to $1:10^6$) with metal abundances varied, too, in the temperature range 11 000–21 000 K with $\log g = 7$ and 8 (Bues 1970), now continued to 9000 K. Besides He and He$^-$ metals must be considered for pressure and absorption. They determine pressure, stratification and flux in models with $T_{\text{eff}} < 14\,000$ K for a hydrogen deficiency of more than a factor of 10^3.

The models were first compared with non-DA white dwarfs in the UBV two color diagram, where these stars occupy the region around the black body line between 8000 and 20 000 K with the DB stars clustering near the position of a black body of 15 000 K. This position is reached for models with a ratio H:He $\leqslant 1:10^4$, $16\,000 \geqslant$ $\geqslant T_{\text{eff}} \geqslant 11\,000$ K. The cooler models form a sequence parallel to the black body line, the effective temperature being two thousand degrees lower than black bodies of equal color.

Up to this point the influence of lines on stratification and flux was neglected. For the structure of the models *blanketing* by HeI lines is less important than that of Balmer and Lyman lines for DA-stars because of the relative weakness of the lines. Metal lines would be more important if metals were present at solar abundances. For the UBV colors, however, *blocking* effects of both, helium and metal lines have to be taken into account for all three filters and thus a temperature determination for DB and DC stars from observed colors becomes difficult.

More suitable for helium-rich stars is the intermediate band system of Strömgren, uvby: the u filter is centered at the steepest absorption edge of HeI in the near UV and affected by HeI lines for models hotter than 14 000 K, the b filter is influenced by the weak HeI line at 4713 Å (max. $W_\lambda \approx 8$ Å for $T_{\text{eff}} = 15\,500$ K), whereas the y filter covers a line-free region. Figure 1 shows the experimental data for 6 DB and

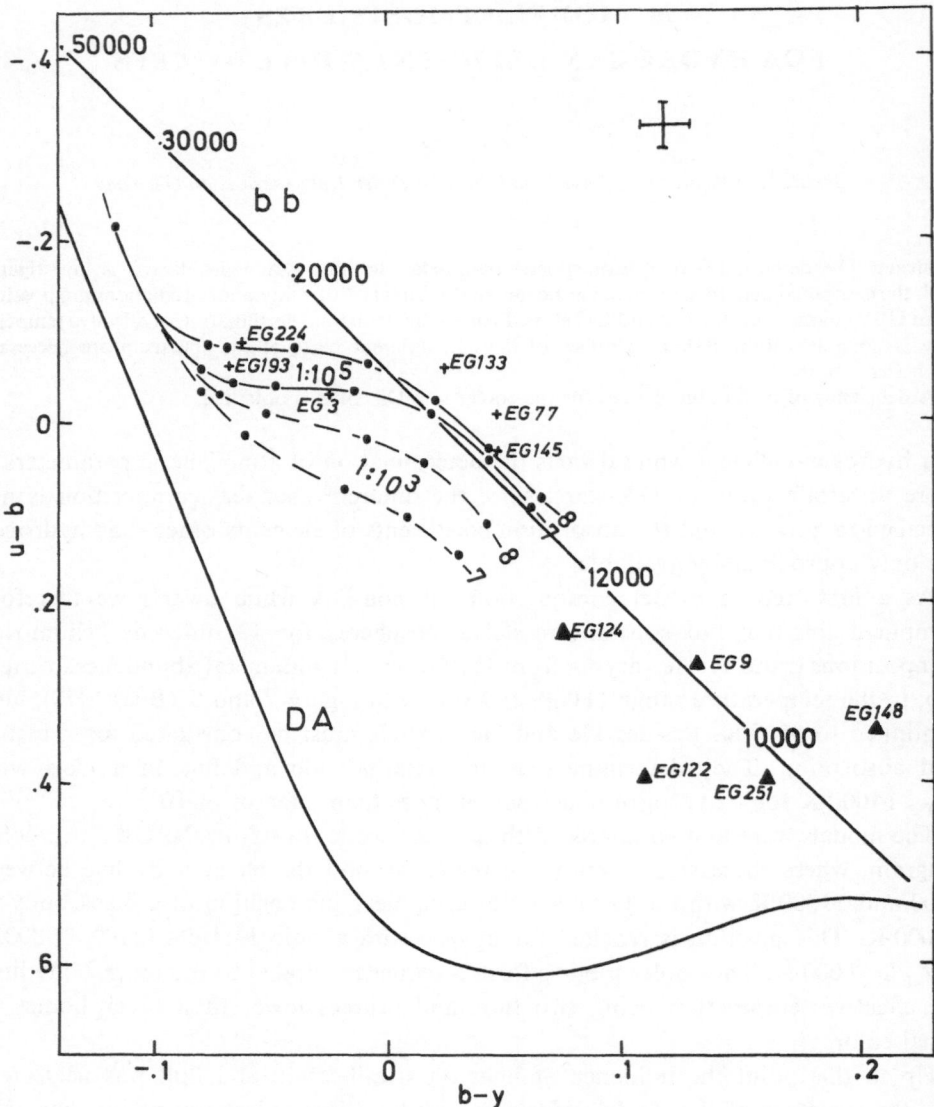

Fig. 1. Strömgren color diagram for white dwarfs. Observed colors for DB(+) and DC(▲) stars compared to models with normal metal abundances, $\log g = 7$ and 8. From right to left $T_{eff} = 10000$, 11000, 12000, 13200, 15500, 18000, 21000 and 30000 K is marked by the dots for H:He = $1:10^5$. For H:He = $1:10^3$ models with $T_{eff} = 10000$ and 21000 K are omitted.

5 DC stars, kindly provided by Graham (1970) and the computed colors for models with solar metal abundances by mass. The error bars indicate the assumed accuracy of measurements. Here we see a well defined sequence of DB stars which does not appear in the UBV two color diagram for the same stars. The observed sequence is reproduced by computed colors for a hydrogen abundance equal or smaller than $1:10^5$. (The position does not change with further reductions of hydrogen.)

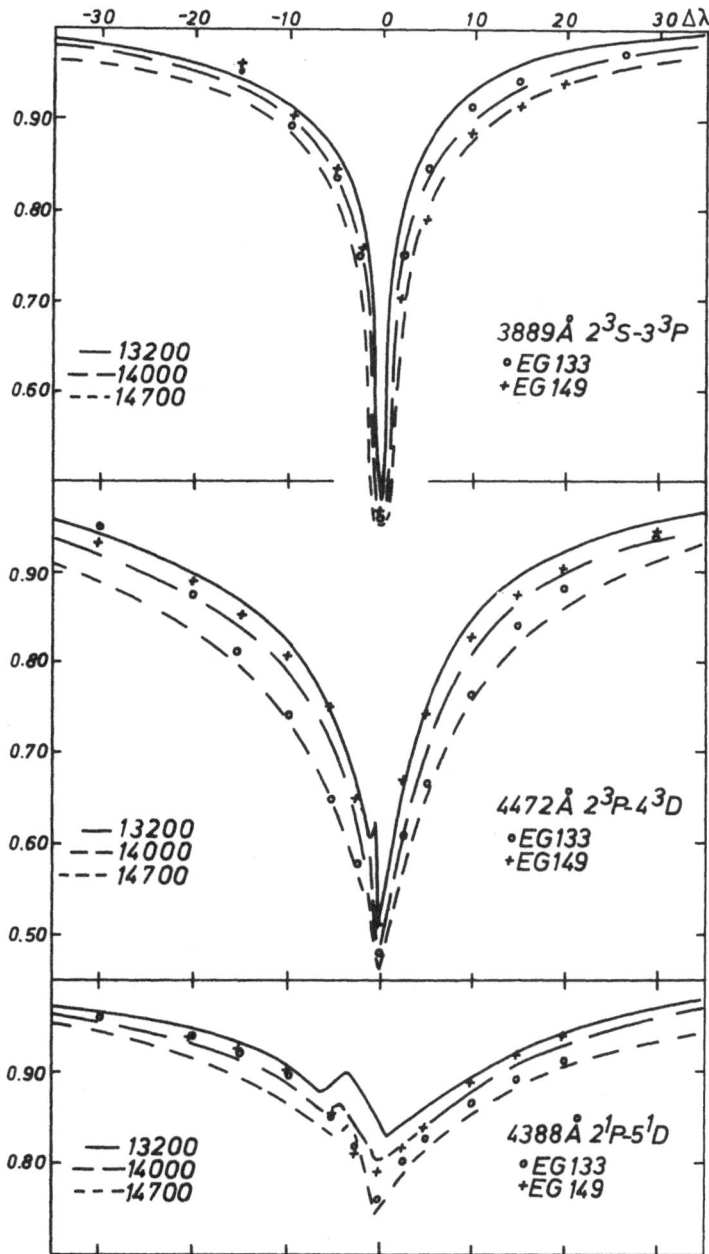

Fig. 2.   Line profiles (in Å) of 3 He I lines for 3 models with composition H:He = 1:10⁵, $Z = 0.02$, $\log g = 7$ compared with observations for EG 133 and EG 149.

Some uncertainties arise for this system from the absolute calibration. The calibration constant obtained by Matsushima (1969) with fluxes of line-blanketed model atmospheres for A stars and ours, which was derived with the flux of δ Del analysed by Reimers (1969), differs by 0.02 mag in $(u-b)$. The corresponding differences in $\log g$ are within the accuracy of computation and temperature determination is possible only within 1000 K for the cooler DB stars.

An improved determination of temperatures is possible by consideration of line profiles. Observed profiles are available for EG 133, EG 149 and EG 145 from Greenstein (1960). The application of broadening theories of Griem et al. (1962), Griem (1969) and Pfennig and Trefftz (1965, 1966) to 10 HeI lines yielded the observed line depths for models with $12\,000 \leqslant T_{\text{eff}} \leqslant 15\,500$ K, $\log g = 7$. A significant difference in line shape (5–25 Å from the center) between $\log g = 7$ and 8 exists for models with $T_{\text{eff}} \leqslant 14\,000$ K only. A comparison of observed and computed profiles for EG 133 and EG 149 is shown in Figure 2 at normal metal abundances.

From relative intensities and shapes of lines in different spectral regions better agreement with observation is obtained if metal abundances are reduced by a factor of 100. The results for these two stars are $T_{\text{eff}} = 14\,700 \pm 500$ K for EG 133 and $T_{\text{eff}} = 14\,000 \pm 500$ K for EG 149, respectively.

An independent check on metal abundances is the calculation of strong metal lines, most important being CaII H and K and MgI $\lambda\lambda 3829$, 3832 and 3838 Å. The equivalent width of the H and K blend for $T_{\text{eff}} = 13\,200$ K, $\log g = 8$ is 30 Å (max ≈ 40 Å for $T_{\text{eff}} = 11\,000$ K) with a sharp line center (central depth 70%) and halfwidths of the order of 10 Å. Thus it must be concluded that Ca is deficient by a factor of at least 100 for DB stars. For Mg, Si, Fe and Ni reduction factors of ~ 10 are necessary in this range of temperature. For lower temperatures, $T_{\text{eff}} \leqslant 11\,000$ K, the computed line strengths increase because of the relatively low He$^-$ absorption coefficient and a DC continuum would not appear for normal metal abundances, although the strongest HeI line in the visible region, $\lambda 4472$ Å, is below visibility. ($W_\lambda \leqslant 0.4$ Å for $T_{\text{eff}} \leqslant 10\,000$ K)

Lines of nitrogen and carbon are too weak to be visible in the spectra.

Since the $\lambda 4670$ stars evidently have carbon, we calculated the partial pressure of C_2 and derived abundances relative to CI at each depth of models with $T_{\text{eff}} = 11\,000$ and 10000 K. With f values of Smith (1969) a strength of 0.1 and 0.5 Å, respectively, for the Swan-band was obtained, whereas the sum of the 6 CI lines $\lambda\lambda 4768$–4775 Å resulted in a total equivalent width of 2 and 6 Å, respectively. This may indicate an effective temperature below 10000 K for the $\lambda 4670$ stars if they have equal composition, since the Swan-band is predominant in their spectra. Vice versa, for the DC stars, we must conclude that carbon is deficient, because the intensities of the CI lines increase for lower effective temperatures and should be visible ($W_\lambda > 6$ Å) for normal abundance.

References

Bues, I.: 1970, *Astron. Astrophys.* **7**, 91.
Graham, J. A.: 1970, private communication.

Greenstein, J. L.: 1960, *Stars and Stellar Systems* 6 (ed. by J. L. Greenstein), University of Chicago Press, Chicago, p. 676.

Griem, H. R.: 1968, *Astrophys. J.* **154**, 1111.

Griem, H. R., Baranger, M., Kolb, A. C., and Oertel, G.: 1962, *Phys. Rev.* **125**, 177.

Matsushima, S.: 1969, *Astrophys. J.* **158**, 1137.

Pfennig, H. and Trefftz, E.: 1965, *Instit. ber. MPI/PAE/Astro.* Nr. 18, Munich.

Pfennig, H. and Trefftz, E.: 1966, *Z. Naturforsch.* **21a**, 697.

Reimers, D.: 1969, *Astron. Astrophys.* **3**, 94.

Smith, W. H.: 1969, *Astrophys. J.* **156**, 791.

19. CONVECTION ZONES AND CORONAE OF WHITE DWARFS

K. H. BÖHM and J. CASSINELLI

Astronomy Dept., University of Washington, Seattle, Wash., U.S.A.

Abstract. Outer convection zones of white dwarfs in the range 5800 K $\leq T_{eff} \leq$ 30000 K have been studied assuming that they have the same chemical composition as determined by Weidemann (1960) for van Maanen 2. Convection is important in all these stars. In white dwarfs $T_{eff} <$ 8000 K the adiabatic temperature gradient is strongly influenced by the pressure ionization of H, He I and He II which occurs within the convection zone. Partial degeneracy is also important.

Convective velocities are very small for cool white dwarfs but they reach considerable values for hotter objects. For a white dwarf of $T_{eff} =$ 30000 K a velocity of 6.05 km/sec and an acoustic flux (generated by the turbulent convection) of 1.5×10^{11} erg cm^{-2} sec^{-1} is reached. The formation of white dwarf coronae is briefly discussed.

During the last 3 years it has become evident that the existence of outer convection zones of cool white dwarfs strongly reduces the core temperatures, the energy contents and consequently the remaining lifetimes of these objects. Preliminary calculations of convection zones have been carried out by van Horn (1970) and by myself (1968, 1969).

It has been pointed out by Greenstein (1969) that the change in the cooling times can become very large because of a feedback between the lowering of the core temperature by convection and the reduction of the specific heat below the Debye temperature.

All earlier calculations of white dwarf convection zones are based on very crude approximations with respect to the thermodynamics. It should be mentioned that in the cool white dwarfs the hydrodynamic aspect does not cause as much trouble as in other stars: The densities are usually so high that one gets an almost sudden transition from a pure radiative to an almost adiabatic gradient.

What kind of improvements are required in the thermodynamics? All previous calculations have used the assumptions that we have a perfect partially ionized gas in the convection zone and that the ionization is governed by the Saha equation.

However, it is clear now that the following complications certainly do arise:

(1) Relatively high (though not complete) electron degeneracy does occur within the convection zones.

(2) H and He I are completely and He II is partially pressure ionized within the convection zone. The usual Saha equation is not an acceptable approximation.

We have now carried out calculations in which we take into account these two complications:

Pressure ionization is treated according to the Stewart and Pyatt (1966) theory. If applied in a naive way this theory leads to a discontinuity in the entropy and to a singularity of the adiabatic gradient at the point where the ionization becomes complete. In order to avoid this problem we have followed a suggestion by Rouse (1964) which permits a rapid but smooth transition to the state of complete pressure ionization.

Luyten (ed.), White Dwarfs, 130–135. All Rights Reserved.

The following effects of partial degeneracy of the electron gas have been taken into account:

(1) The equation of state has been determined assuming that the atoms and ions are nondegenerate but permitting a partial degeneracy of the electrons.

(2) The influence of this degeneracy on the entropy S, the adiabatic gradient ∇_{ad}, the scale height H has been taken into account.

(3) In the mixing length theory one assumes usually that temperature and density fluctuations are related according to

$$\delta \varrho / \varrho = - \delta T / T \tag{1}$$

as follows from the assumption of constant pressure within a horizontal plane. We have written

$$\delta \varrho / \varrho = - D \left(\delta T / T \right) \tag{2}$$

where D is now a complicated factor depending on T, ϱ, the degeneracy parameter α and the Fermi Dirac functions.

Numerical calculations have been carried out for white dwarfs of the Weidemann chemical composition (i.e. consisting mostly of He, with 2% H and a considerable underabundance of heavier elements) in the range 5800 K $\leqslant T_{eff} \leqslant$ 30000 K. We have used a surface gravity of 10^8 which corresponds to white dwarfs of 0.6 M according to the Chandrasekhar (1939) mass-radius relation.

One of the important results of these calculations is the following: For white dwarfs with $T_{eff} < 8000$ K the adiabatic gradient is strongly influenced by the presence of the three pressure ionization zones for H, HeI and HeII. In these regions the adiabatic gradient drops to very low values. The net effect of this behavior is of course a lowering of the average adiabatic gradient.

This leads (as compared to the earlier simple models) to a higher degree of degeneracy for a given temperature at the bottom of the convection zone. Consequently we get an even lower core temperature than predicted earlier. A relatively small part of this effect is compensated by the influence of partial degeneracy.

The effect is illustrated in Figure 1 which show the depth dependence of the adiabatic and the radiative gradients in a white dwarf convection zone. We can clearly see the three distinct pressure ionization zones for H, HeI and HeII. In a star like van Maanen 2 we get two separate convection zones the lower of which reaches a density of 8.015×10^2 g cm^{-3} and degeneracy parameter $\alpha = -\mu/kT$ of -13.999.

It is interesting to note that all white dwarfs up to an effective temperature of 30000 K have an outer convection zone. This means that convection should be taken into account in practically all studies of white dwarf atmospheres and of the internal thermal structure of these stars.

Figures 2, 3 and 4 show the depth of the convection zone, the temperature and the degeneracy parameter α at the lower boundary of the zone. These figures indicate that convection is much more important for the thermal structure of relatively cool white dwarfs than for hot white dwarfs. This conclusion is based on the following results:

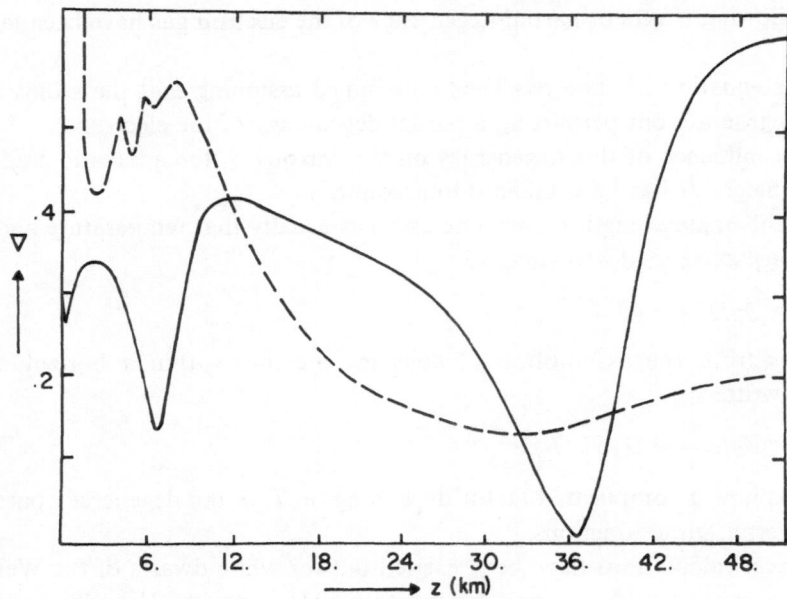

Fig. 1.   The adiabatic (solid curve) and the radiative (broken line) gradient as a function of geometrical depth for a white dwarf with $T_{\text{eff}} = 6000$ K, $g = 10^8$ cm sec^{-2}. (From Böhm, 1970.)

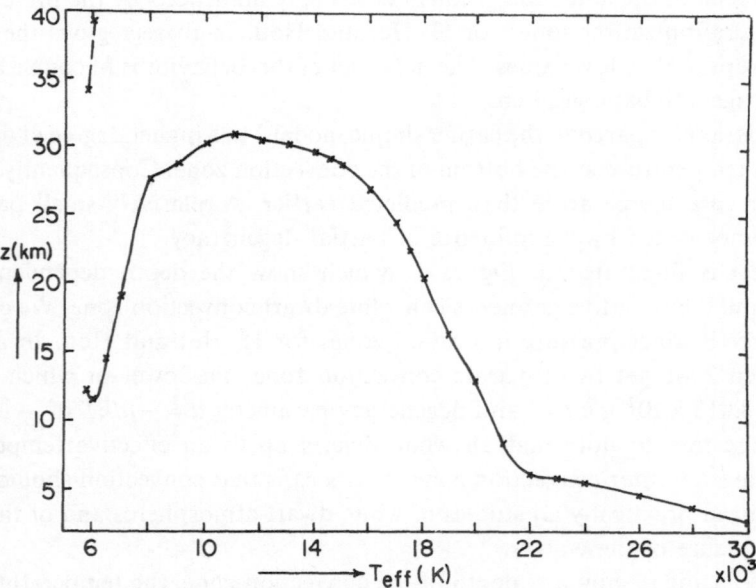

Fig. 2.   The depth of the convection zone as a function of effective temperature. (The crosses correspond to the models which have been computed.) $g = 10^8$ cm sec^{-2} has been used for all models. The short broken line corresponds to the separate lower convection zone which occurs in cool white dwarfs.

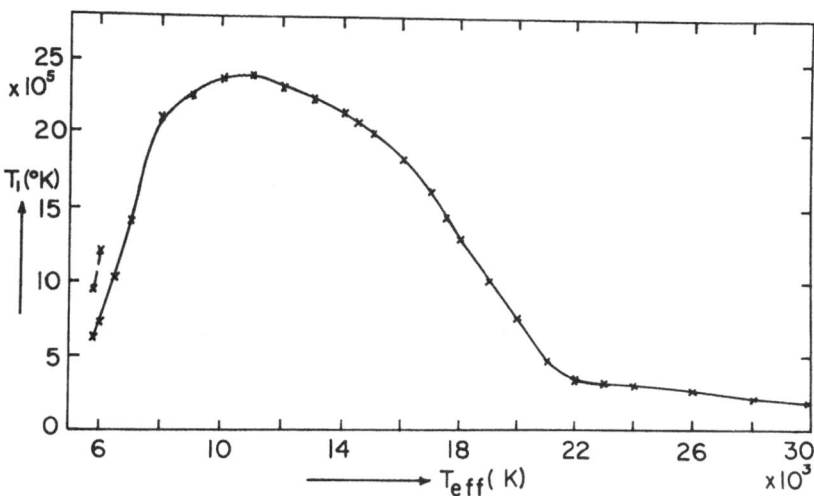

Fig. 3. The temperature at the lower boundary of the convection zone for white dwarfs of different effective temperatures. (The broken line has the same meaning as in Figure 2.)

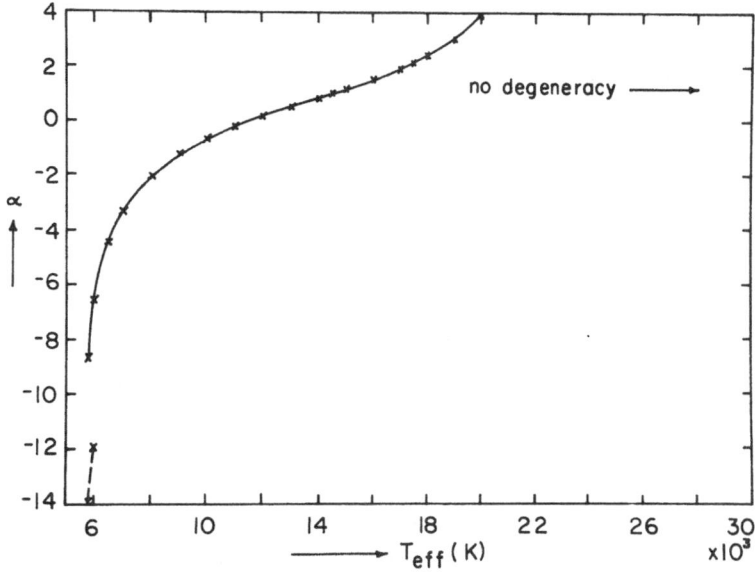

Fig. 4. The degeneracy parameter $\alpha = -\mu/kT$ at the lower boundary of the convection zone for white dwarfs of different effective temperatures. (The broken line has the same meaning as in Figure 2.)

(1) Beyond $T_{eff} = 11\,000$ K the thickness of the convection zone starts to decrease with increasing T_{eff}. For high T_{eff} the outer convection zones become relatively thin.

(2) The degeneracy parameter $(-\alpha)$ at the bottom of the convection zone decreases rapidly with increasing T_{eff}. This means that for cool white dwarfs the convection zone reaches down to the 'transition point' (see e.g. Schwarzschild, 1958) to the almost

isothermal core and consequently directly influences the temperature gradient in the whole nondegenerate envelope. In a hot white dwarf on the other hand the convection zone covers only a small fraction of the distance to the transition point.

One intriguing result of our computations is related to the convective velocities. As one might have expected the velocities are very low in cool white dwarfs. On the other hand, in hot white dwarfs with their much lower densities and their higher energy fluxes the convective velocities can become considerable. A value of 6.05 km/sec is reached in a white dwarf with an effective temperature of 30000 K. The velocity of sound at the same point in the same convection zone is 18.80 km/sec. This leads to a Mach number M of 0.322 which is higher than the maximal Mach number occurring in the solar convection zone.

If we use the usual estimate of the acoustic energy flux

$$H_{ac} = 19\varrho v_{\max}^3\, M^5 \tag{3}$$

(Lighthill, 1952; Bierman and Lüst, 1960) we find an acoustic energy flux of 1.5×10^{11} erg cm^{-2} sec^{-1} which has to be compared to 4.6×10^{13} erg cm^{-2} sec^{-1} for the total flux. These numbers show that hot white dwarfs have a considerable acoustic output and might easily develop chromospheres and coronae. In fact the acoustic flux in such a white dwarf is about twice as high as the total radiative flux in the sun. In view of the high gravitational field of the white dwarf we have to expect a corona which has a higher temperature than the solar corona. This follows from an estimate as suggested by Shklovsky (1965) who assumes that $(v_{\text{escape}}/v_{\text{thermal}})^2$ is the same for all stellar coronae. From such an estimate we get temperatures of the order of 10^8 K for white dwarf coronae. A qualitative extrapolation of Kuperus' (1965) results to stars with high g indicates a temperature of only a few million degrees. It is interesting to note that such a hot white dwarf with a corona could have a relatively high ratio of X-ray emission to optical emission because of

(1) the relatively high ratio of acoustic to total flux in these objects

(2) the high bolometric correction which is required even if one neglects the radiation of the corona.

However, the X-ray radiation certainly cannot be strong enough to explain strong thermal X-ray sources like Sco X-1.

A further question which cannot yet be answered is whether some of the emission lines observed in some white dwarfs (see e.g. Greenstein, 1958, 1960) can be attributed to chromospheres heated by acoustic waves.

References

Bierman, L. and Lüst, R.: 1960, in *Stellar Atmospheres* (ed. by J. L. Greenstein), University of Chicago Press, Chicago.

Böhm, K. H.: 1968, *Astrophys. Space Sci.* **2**, 375.

Böhm, K. H.: 1969, in *Low Luminosity Stars* (ed. by S. S. Kumar), Gordon and Breach, New York, p. 393.

Böhm, K. H.: 1970, *Astrophys. J.* **162**, 919.

Chandrasekhar, S.: 1939, *An Introduction to the Study of Stellar Structure*, University of Chicago Press, Chicago.

Greenstein, J. L.: 1958, in *Handbuch der Physik* **50** (ed. by S. Flügge), Springer-Verlag, Berlin, p. 161.

Greenstein, J. L.: 1960, in *Stellar Atmospheres* (ed. by J. L. Greenstein), University of Chicago Press, Chicago, p. 676.

Greenstein, J. L.: 1969, *Comments Astrophys. Space Phys.* **1**, 62.

Kuperus, M.: 1965, *The Transfer of Mechanical Energy in the Sun and the Heating of the Corona*, Reidel, Dordrecht.

Lighthill, M. J.: 1952, *Proc. Roy. Soc. (London)* **211**, 564.

Rouse, C. A.: 1964, *Astrophys. J.* **139**, 339.

Schwarzschild, M.: 1958, *Structure and Evolution of the Stars*, Princeton University Press, Princeton.

Shklovskij, I. S.: 1965, *Physics of the Solar Corona*, Pergamon Press, Oxford.

Stewart, T. C. and Pyatt, K. D.: 1966, *Astrophys. J.* **144**, 1203.

Van Horn, H. M.: 1970, *Astrophys. J. (Letters)* **160**, 53.

Weidemann, V.: 1960, *Astrophys. J.* **131**, 638.

20. GRAVITATIONAL SORTING AND OVERSTABILITY
IN WHITE DWARFS

A. BAGLIN

Observatoire de Nice, Nice, France

1. Introduction

Zones with μ gradient in stars have been discussed extensively in recent years in relation with the so-called 'semi-convective zone' which forms in massive stars when the convective core shrinks.

When the condition $\nabla_T \in [\nabla_{Ta}, \nabla_{Ta} + (\beta/4 - 3\mu) \nabla_\mu]$ is satisfied (β is the classical ratio P_{Gas}/P_{Total}, ∇_{Ta} and ∇_T the adiabatic and actual temperature gradients, $\nabla\mu$ the molecular weight gradient) motions are locally overstable.

A local treatment of this zone led Kato (1966) to the conclusion that these overstable motions produce an important mixing and that the gradient remains almost adiabatic.

Then Gabriel (1969) and Auré (1970) tried to answer the question whether motions can settle in the whole star taking into account the stabilizing effect of the large radiative regions above and below the semi convective zone. Due to the strong stabilization of the outer radiative zones Auré concluded that no motions can be maintained in the star and then no mixing will be achieved.

The most favorable situation to excite overstability of the whole star would be when the zone of varying molecular weight lies very close to the surface, where the eigenfunction is large. Such a situation could be achieved in white dwarfs. It has been shown by Schatzman (1958) that large gravity can lead to a sorting of the elements and to mild thin zones of varying chemical composition. In the case of a white dwarf with a very thin hydrogen envelope, the zone where hydrogen and helium separate can be very superficial. If the star is not too hot the condition $\nabla_{Ta} < \nabla_T$ will be satisfied. The purpose of this paper is to examine the influence of this zone on the stability of the star.

2. The Zone of Sorting of Helium in the Envelopes of White Dwarfs

We first estimate the extend of the sorting zone and then the speed of sorting. We will assume that the hydrogen-helium transition occurs high in the envelope, the characteristic parameters being:

$$\log T = 4.12$$
$$\log P = 6$$

These parameters have been estimated from hydrogen rich envelope models constructed by Vauclair (1970).

For the sake of simplicity we assume that only fully ionized hydrogen and helium

Luyten (ed.), White Dwarfs, 136–144. All Rights Reserved.

are present. Let us call Z_1, n_1, m_1, the charge, number density and mass of hydrogen ions; Z, n, m the same quantities for helium ions, e_1, s the charge and number density of electrons. We will use the u and v variables defined by:

$$\frac{dv}{v} = -\frac{mg}{kT} dz$$

$$\frac{dv}{v} = \frac{e\psi'}{kT} dz \tag{1}$$

where g is the acceleration of gravity and ψ' the electrical potential.

The statistical equilibrium is:

$$n = n_0 \frac{T_0}{T} u^4 v^2$$

$$n_1 = n_{10} \frac{T_0}{T} uv \tag{2}$$

$$s = s_0 \frac{T_0}{T} u^{m_e/m_1} v^{-1} \simeq s_0 \frac{T_0}{T} v^{-1}$$

n_0, n_{10}, s_0 T_0, being the number density of helium, hydrogen and electrons, and the temperature at the same reference level.

The extend of the zone of sorting is given by

$$\frac{1}{n}\frac{du}{dz} = -\frac{1}{T}\frac{dT}{dz} + \frac{4}{u}\frac{du}{dz} + \frac{2}{v}\frac{dv}{dz} \tag{3}$$

$$\frac{1}{u}\frac{du}{dz} = -\frac{1}{R}\frac{1}{\varrho T}\frac{dP}{dz} = -\mu\frac{1}{P}\frac{dP}{dz} \tag{4}$$

The estimate of $1/v\ dv/dz$ is obtained using the condition of electrical neutrality which relates u and v.

$$n_1 Z_1 + nZ = s \tag{5}$$

At the reference level one has

$$n_{10} Z_1 + n_0 Z = s_0 \tag{6}$$

and at any level

$$2\frac{n_0}{s_0} u^4 v^3 + \frac{n_{10}}{s_0} uv^2 = 1 \tag{7}$$

If one chooses the reference level as being the level of half sorting

$$n_{10} = 0.5 \frac{\varrho}{m_1} \qquad n_0 = 0.5 \frac{\varrho}{4m_1}$$

$$\frac{n_0}{s_0} = \frac{1}{6} \quad \text{and} \quad \frac{n_{10}}{s_0} = \frac{2}{3} \tag{8}$$

Equation (7) becomes

$$u^4 v^3 + 2uv^2 - 3 = 0 \qquad (9)$$

So that $1/v \; dv/dz$ can be obtained by

$$\frac{1}{v}\frac{dv}{dz} = -\frac{1}{u}\frac{du}{dz}\left(\frac{4u^3 v^3 + 2v^2}{3u^3 v^2 + 4u}\right) \qquad (10)$$

In the vicinity of $u = v = 1$ this gives

$$\frac{1}{v}\frac{dv}{dz} \simeq -\frac{6}{7}\frac{1}{u}\frac{du}{dz} \qquad (11)$$

Then Equation (3) is

$$\frac{1}{n_1}\frac{dn_1}{dz} = -\left(\nabla_T + \frac{16}{7}\mu\right)\frac{1}{p}\frac{dp}{dz} \qquad (12)$$

For a typical white dwarf atmosphere

$$\left|\frac{1}{u_1}\frac{du_1}{dz}\right| \simeq 2\left|\frac{1}{p}\frac{dp}{dz}\right| \qquad (13)$$

So that the extent of the zone of sorting is of the order of half a pressure scale height i.e. 2 temperature scale heights. In the typical case, this leads to a distance of approximately 5×10^4 cm.

The velocity of the elements, as given by Chapman and Cowling (1960) is

$$\begin{aligned}
v &= \lambda_D k_1 \frac{1}{p}\frac{dp}{dz} \\
k_1 &= 2A - i - 1 \\
\lambda_D &= 6.62 \times 10^9 \; T^{5/2}/(n_2 M_2 i\alpha)
\end{aligned} \qquad (14)$$

A, i, being the molecular weight and the degree of ionisation of the sorted element; α is a slowly varying factor of the order of 10.

$$M_2 = \frac{m}{m_1 + m}$$

with the typical values chosen here

$$V \simeq 10^{-3} \text{ cm/sec}$$

So that the sorting is achieved in 5×10^7 sec.

If no mixing processes like turbulence occur in the envelope the sorting is completely achieved during evolution.

The corresponding gradient of molecular weight is of the order of

$$\nabla\mu = \frac{d\log\mu}{d\log p} \simeq 0.1$$

The extension of the zone $\nabla_T < \nabla_{Ta}$ in hydrogen rich envelopes of white dwarfs is extremely sensitive to the effective temperature and less sensitive to gravity.

In a typical model $M = 0.88\, M_\odot$ $L = 0.003\, L_\odot$ $\log T_{\text{eff}} = 4.12$, $\log p$ is in the range [5.44; 6.59] and $\log T$ [4.03; 4.45]. Smaller effective temperature leads to strongly convectively unstable models, and larger effective temperature to strongly radiatively stable ones.

3. Stability of the Gravity Modes

As the interesting region lies in the very outer layers of the star we assume a plane parallel medium stratified along the z axis. Equations of continuity, motion, radiative equilibrium (in the diffusion approximation) and energy lead to the well known system for the displacement and the perturbations:

$$
\begin{aligned}
&\nabla \cdot r' + \frac{\varrho'}{\varrho} + z' \frac{\partial \log \varrho}{\partial z} = 0 \\[6pt]
&\sigma^2 r' = \frac{1}{\varrho} \nabla p' + g \frac{\varrho'}{\varrho} \\[6pt]
&\nabla \left(\frac{4}{3} a T^3 T' \right) = -\frac{\kappa \varrho}{c} F' - \left(\frac{\kappa \varrho}{c} \right)' F \\[6pt]
&\left(\frac{p'}{p} + z' \frac{\partial \log p}{\partial z} \right) - \nabla_{ad} \left(\frac{\varrho'}{\varrho} + z' \frac{\partial \log \varrho}{\partial z} \right) = -\frac{\nabla_{pa} - 1}{i \sigma p} (\nabla \cdot F)'
\end{aligned}
\tag{15}
$$

r' is the displacement and z' its component along the z axis; ϱ, p, T density, pressure, temperature; g gravity, along the z axis; F, χ radiative flux and Rosseland mean opacity.

All perturbed quantities are written in the form

$$
A = A_0 + A'(z)\, e^{i\sigma t + k_x x + k_y y}
\tag{16}
$$

A_0 is the equilibrium value and A' the eulerian perturbation, function of z only.

We also assume the perfect gas law, with no contribution of the radiative pressure:

$$
P = R \varrho T / \mu
\tag{17}
$$

μ the molecular weight is assumed to be constant along the motion, the diffusion time being much longer than the period of the oscillations; so that

$$
\mu'/\mu + z'\, \partial \log \mu / dz = 0
\tag{18}
$$

As usual we will carry a linear analysis and the system (15) can be written in terms of the variable $\omega = \log p$ and the reduced functions

$$
\xi = z'/H_p, \quad \theta = T'/T, \quad \pi = p'/p, \quad \varphi = F_z'/F
$$

$$
\frac{\partial \xi}{\partial \omega} = (\nabla \mu - 1)\, \xi + \left(1 - \frac{k^2 g H_p}{\sigma^2} \right) \pi - \theta
\tag{19}
$$

$$\frac{\partial \pi}{\partial \omega} = \left(\nabla \mu - \frac{\sigma^2 H_p}{g}\right) \xi - \theta \tag{20}$$

$$\frac{\partial \theta}{\partial \omega} = (1 + \kappa_\mu) \nabla_\mu \nabla_T \xi + (1 + \kappa_p) \nabla_T \pi + (\chi_T - 5) \nabla_T \theta + \nabla_T \varphi \tag{21}$$

$$\frac{\partial \varphi}{\partial \omega} = -\frac{i\sigma p H_p}{F \nabla_{Ta}} \left[(\nabla_T - \nabla_{Ta}) \xi + (1 + \nabla_{\varrho a}) \pi - \theta\right] + \frac{k^2 H_p^2}{\nabla_T} \theta \tag{22}$$

In atmospheres of white dwarfs the three quantities $k^2 g H/\sigma^2$, $\sigma^2 H_p/g$ and $k^2 H_p^2/\nabla_T$ are very small and will be neglected all along.

Auré (1970) has shown that due to the exponential increase of the adiabatic eigenfunction close to the surface the influence of the radiative layers situated below the region of varying molecular weight is negligeable. But a large damping comes from the outer layers. We will assume here that the region of sorting coincides with the photosphere. This means that the star is composed almost entirely of helium and keeps only a thin atmosphere of hydrogen.

To be able to proceed further in the analytical calculations we will make two drastic approximations. In the region of varying molecular weight extending from z_1 to z_0 motions are adiabatic, although in the outer layers from z_0 to z_2 they are isothermal.

The actual eigenfunction is $\xi = \xi_0 \, e^{i\sigma t + \sigma' t}$ and the damping coefficient is expressed by (Ledoux and Walraven, 1958)

$$\sigma' = \frac{\int \frac{\delta T}{T} (\nabla \cdot F)' \, dv}{\int \varrho r' \cdot r' \, dv} \tag{23}$$

Let us call

$$I_1 = \int_{z_1}^{z_0} \frac{\delta T}{T} \nabla \cdot F' \, dv \quad \text{and} \quad I_2 = \int_{z_0}^{z_2} \frac{\delta T}{T} \nabla \cdot F' \, dv$$

As the lower zone is unstable I_1 is negative; but I_2 is positive. The whole model will be unstable if $I_1 + I_2 < 0$.

4. Excitation of the Sorting Layer

In the adiabatic approximation the eigenfunction can be obtained solving the set of equations:

$$\frac{\partial \xi_0}{\partial \omega} = (\nabla \mu - 1) \xi_0 + \pi - \theta \tag{24}$$

$$\frac{\partial \pi}{\partial \omega} = \nabla \mu \xi_0 - \theta \tag{25}$$

$$(\nabla_T - \nabla_{Ta}) \xi_0 + (1 - \nabla_{\varrho a}) \pi - \theta = 0 \tag{26}$$

We assume that in the layer the different gradients are constant. The second order differential equation giving ξ_0 is

$$\ddot{\xi}_0 + (2 - \nabla_\varrho)\, \dot{\xi}_0 + (\lambda - \nabla_\varrho)\, \xi_0 = 0 \tag{27}$$

And the general solution can be written

$$\xi_0 = A \exp(-\omega)[1 + \exp(\nabla_\varrho \omega + \alpha)] \tag{28}$$

We set the upper boundary of the layer at $\omega = 0$ and assume $\xi_0 = 1$ i.e. $\alpha = 0$, $A = \tfrac{1}{2}$. And the eigenfunction is

$$\xi_0 = \tfrac{1}{2} e^{-\omega}[1 + e^{\nabla_\varrho \omega}] \tag{29}$$

In the quasi adiabatic approximation $\delta T/T = \theta - \xi \nabla_T$ using (26) becomes

$$\delta T/T = \nabla_{Ta}(\pi - \xi) \tag{30}$$

and $\nabla \cdot F'$ is computed from (21).

The adiabatic expression for π is derived from (24)

$$\pi = - \frac{\nabla_\varrho - \nabla_{\varrho a}}{2\nabla_{\varrho a}} e^{-\omega} + \tfrac{1}{2} e^{(\nabla_\varrho - 1)\omega} \tag{31}$$

Then

$$\nabla \cdot F' = \frac{F}{H_p}\left\{\left(-a + b\,\frac{\nabla_\varrho}{\nabla_{\varrho a}}\right) e^{-\omega} + a\,(\nabla_\varrho - 1)\, e^{(\nabla_\varrho - 1)\omega}\right\} \tag{32}$$

where a and b are expressed in terms of characteristic quantities of the layer

$$a = -1 - \kappa_\mu \nabla \mu - \kappa_p - (\kappa_T - 4)\,\nabla_T$$

$$b = -\frac{\nabla_{Tq}}{\nabla_T} - \kappa_p - (\kappa_T - 4)\,\nabla_{Tq}$$

χ_μ, χ_T, χ_p being the logarithmic derivatives of χ with respect to μ, T and p. The contribution of this layer to the damping integral is:

$$I_1 = -\int_{\omega_1}^{0} \frac{\delta T}{T}\,\nabla \cdot F\, H_p\, d\omega \tag{33}$$

Performing the integration one obtains

$$I_1 = F\nabla_{Ta}\,\frac{\nabla_\varrho}{2\nabla_{\varrho a}}\,[g(\omega_1) - g(0)] \tag{34}$$

where

$$\omega_1 = \log\frac{p_1}{p_0} \quad \text{and} \quad g(\omega) = \tfrac{1}{2}\left(-a + b\,\frac{\nabla_\varrho}{\nabla_{\varrho a}}\right) e^{-2\omega} - \frac{a\,(\nabla_\varrho - 1)}{\nabla_\varrho - 2}\, e^{(\nabla_\varrho - 2)\omega}$$

5. Damping by the Outer Layers

The boundary of a possible semi convective zone is situated in the very outerlayers ($\tau \simeq 0.2$) so that it is impossible to keep the quasi adiabatic approximation in this region.

On the other hand, as the temperature relaxation time scale is very short one can assume that the perturbation has always the same temperature as the medium i.e. $\theta = 0$. The eigenfunction can be computed from

$$\frac{\partial \xi_0}{\partial \omega} = -\xi_0 + \pi$$
$$\frac{\partial \pi}{\partial \omega} = -\frac{\sigma^2 H_p}{g} \xi_0 \tag{35}$$

As $\sigma^2 H_p / g \ll 1$ the solution is

$$\xi a = K_1 e^{-\omega} + K_2 e^{-(\sigma^2 H_P/g)\,\omega}$$
$$\pi = K_2 e^{-(\sigma^2 H_P/g)\,\omega} \tag{36}$$

Boundary conditions at $\omega = 0$ fix the constants. They are

$$\xi_1(0) = \xi_2(0)$$
$$\pi_1(0) = \pi_2(0)$$

where index 1 and 2 refer to the lower and upper zones respectively

$$K_1 + K_2 = 1$$
$$K_2 = \tfrac{1}{2} - \frac{\nabla_\varrho - \nabla_{\varrho a}}{2}$$
$$K_2 = \frac{\nabla_\varrho}{2\nabla_{\varrho a}} \quad K_2 = 1 - \frac{\nabla_\varrho}{2\nabla_{\varrho a}} \tag{37}$$

$\nabla \cdot F'$ is computed using (22) with $\theta = 0$

$$(\nabla \cdot F)' = \frac{F}{H} \frac{\partial \varphi}{\partial \omega} = -\frac{i\sigma p}{\nabla_{Ta}} \left[(\nabla_T - \nabla_{Ta}) \xi + (1 - \nabla_{\varrho} a) \pi \right] \tag{38}$$

keeping only the first term in $e^{-\omega}$

$$\nabla \cdot F = -\frac{i\sigma p}{\nabla_{Ta}} (\nabla_T - \nabla_{Ta}) K_1 e^{-\omega} \tag{39}$$

$\delta T/T$ is estimated from Equation (21) where $\nabla_\mu = 0$ and $\theta = 0$

$$\partial \theta / \partial \omega = (1 + \chi_p) \nabla_T \pi + \nabla_T \varphi \tag{40}$$

This equation is the diffusing approximation of the radiative transfer equation. In the outer layers it can give only rough orders of magnitude.

As $\nabla \cdot F'$ is in quadrature with ξ and π, we only need the part which is in quadrature with ξ in $\delta T/T$ i.e.

$$(\delta T/T)^* = \theta^* \quad \text{with} \quad \partial\theta^*/\partial\omega = \nabla_T \varphi \tag{41}$$

Using the expression of $\partial\varphi/\partial\omega$ from Equation (22) and performing two successive integrations ($\varphi = \theta^* = 0$ at $p = 0$) one gets

$$\theta^* = \frac{i\sigma}{F} \frac{\nabla_T - \nabla_{Ta}}{\nabla_{Ta}} \frac{\nabla_T}{(1 - \nabla_\varrho)^2} K_1 p_0 H_p \tag{42}$$

and then the contribution of this region to the damping integral is

$$I_2 = \int_{z_0}^{z_2} \frac{\sigma^2}{F} \frac{(\nabla_T - \nabla_{Ta})^2}{\nabla_{Ta}^2} \frac{\nabla_T}{(1 - \nabla_\varrho)^2} \left(1 - \frac{\nabla_\varrho}{2\nabla_{\varrho a}}\right)^2 H_p p_0^2 \, dz \tag{43}$$

i.e.

$$I_2 = -\frac{\sigma^2}{F} \frac{(\nabla_T - \nabla_{Ta})^2}{\nabla_{Ta}^2} \frac{\nabla_T}{(1 - \nabla_\varrho)^2} \left(1 - \frac{\nabla_\varrho}{2\nabla_{\varrho a}}\right)^2 H_p^2 p_0^2 \omega_2 \tag{44}$$

6. Numerical Estimates for I_1 and I_2

The typical model gives

$$F = 10^{11.1}$$
$$\sigma \simeq 10^{-1}$$
$$H_p \simeq 10^4 \text{ cm}$$
$$p_0 = 10^{5.5} \text{ cgs}$$

We will stop the integration in the outer atmosphere at $\omega_2 = -3$. This boundary is completely arbitrary but one sees that I_2 is proportional to ω_2 and so not very sensitive to the position of this limit. In the outer layer the mean values of the gradients are:

$$\nabla_T \simeq 0.1$$
$$\nabla_{Ta} \simeq 0.04$$

So that $I_2 \simeq 10^7$

In the region of varying molecular weight, we assume a Kramer's opacity law and the model gives the following mean values

$$\nabla\mu = 0.1$$
$$\nabla_{Ta} = 0.4$$
$$\nabla_\varrho = 0.8$$
$$\omega_1 = 1.3$$

So that $g(\omega_1) - g(0) = -0.32$ and $I_1 \simeq -10^8$ so that $I_1 + I_2 < 0$.

7. Conclusion

The result obtained here that the instability of the whole star could be triggered by the zone of varying molecular weight is only an indication that such a mechanism could work actually.

Very rough approximations are necessary to carry all along analytical expressions. A treatment including a better equation of transfer in the outer layers and a complete resolution of the differential system is needed to conclude. The region where $\nabla_{Ta} < \nabla_T$ coincides with the region of second ionization of helium and we have omitted completely to take into account the variation of the degree of ionisation.

From the expressions obtained from I_1 and I_2 one sees that high order modes are more favorable to instability. If the effective temperature increases H_p^2 / F^2 decreases and instability is favored but on the other hand the radiative gradient remains too small and the condition $\nabla_{Ta} < \nabla_T$ is never satisfied. So that only a small interval in T is favorable to that process.

It also implies that the zone of sorting is very close to the surface which means that there is very little hydrogen left.

References

Auré, J. L.: 1970, *Astron. Astrophys.*, in press.
Chapman, S. and Cowling, T. G.: 1960, *The Mathematical Theory of Non-Uniform Gases*, Reprinted Dover, New York.
Gabriel, M.: 1969, *Astron. Astrophys.* **1**, 321.
Kato, S.: 1966, *PASJ* **18**, 374.
Ledoux, P. and Walraven, T.: 1958, *Handbuch der Physik* **51**, Springer-Verlag.
Schatzman, E.: 1958, *White Dwarfs*, North-Holland Publ. Co., Amsterdam.
Vauclair, G.: 1970, to be published.

21. MODELS OF WHITE DWARFS, RADIAL PULSATIONS AND VIBRATIONAL STABILITY

G. VAUCLAIR

Institut d'Astrophysique de Paris, France

1. Models of White Dwarfs

We have computed a set of models of white dwarfs. The equation of state is the Chandrasekhar (1939) one corrected by Salpeter (1961). It takes into account both the interactions between particles and the effect of non-zero temperature.

The models consist of an isothermal core with 50% of carbon and 50% of oxygen, surrounded by an envelope of population I composition with 70% of hydrogen and 3% of metals by mass. Nuclear energy is liberated by p-p reactions located in a shell at the edge of the isothermal core. Equilibrium models of various masses and luminosities have been built; Table I gives the values of their characteristic parameters. The central temperature-luminosity-mass relation for our models is compared with the Schatzman (1952) one for pure hydrogen envelopes (S), the Hubbard and Wagner (1970) one for Mg core and solar composition envelope models (HW) and the Van Horn (1970) one for carbons models (VH) (Figure 1). At high luminosities there is a sensible departure from the linear relation. Though our models are not evolutionary sequence ones, they fit very well with the sequences calculated by Vila (1966, 1967). The models studied here are of the type expected from the calculation of the evolution of some double systems (Lauterborn, 1970) after the white dwarfs have lost their very diluted envelopes.

2. Eigenvalues and Eigenfunctions of the Radial Pulsations

The eigenvalues and eigenfunctions of radial pulsations are calculated in the adiabatic assumption for the fundamental mode. We confirm that the eigenfunctions do not decrease very much through the white dwarfs. However, it is only for the more massive white dwarfs that the eigenfunction is constant; but for less massive models it decreases inward (Figure 2). No steep gradient are found in the envelopes. The periods vary in the range 3.8 sec for the 1.2 $M_\odot$ models to 18.5 sec for the 0.4 $M_\odot$ one. Our periods are compared with (1) the periods of Harper and Rose's (1969) zero temperature models, (2) the periods of Ostriker and Tassoul's (1969) non rotating white dwarfs models, and (3) the periods of Cohen *et al.* (1969) carbon models; the two first ones built up according to the Chandrasekhar equation of state, and the last one according to the same equation of state improved by Salpeter's corrections (Table II). The decrease of the periods is due to the improvement of both the equations of state and the adiabatic gradient.

Luyten (ed.), White Dwarfs, 145–154. All Rights Reserved.

1	2	3	4	5	6	7	8
No.	$M/M_\odot$	$R/R_\odot$	$L/L_\odot$	$\mathrm{Log}\,g$	$\mathrm{Log}\,T_{\mathrm{eff}}$	M_B	T_c (10^6K)
1	1.2	5.411×10^{-3}	8.5518×10^{-3}	9.050	4.3798	9.8898	9.4107
2	1.2	5.4663×10^{-3}	9.1305×10^{-2}	9.041	4.6347	7.3187	18.1631
3	1	8.0114×10^{-3}	1.0735×10^{-1}	8.63	4.5693	7.1429	17.9109
4	1	7.8694×10^{-3}	9.1288×10^{-3}	8.644	4.3056	9.8189	9.6652
5	0.8	1.0596×10^{-2}	2.0348×10^{-1}	8.293	4.5780	6.4486	19.0456
6	0.6	1.3997×10^{-2}	4.2644×10^{-1}	7.924	4.5979	5.6453	20.2989
7	0.6	1.2763×10^{-2}	2.1026×10^{-3}	7.994	4.0412	11.4131	6.6209
8	0.4	1.6942×10^{-2}	3.7687×10^{-3}	7.587	4.0430	10.7795	8.00

Notes – 1: number of the model; 2: mass in solar mass unit; 3: radius in solar radius unit; 4: luminosity magnitude; 8: central temperature in million degree K; 9: central pressure in c.g.s.; 10: central density in g cm shell; 13: fraction of the radius occupied by the partially degenerate shell; 14: fraction of the radius occup

The effect of the temperature is an increase of the radius and of the pulsation period. For instance, the first $0.6\,M_\odot$ model, number 7, with a central temperature of 6.6×10^6 K, has a radius of 8.96×10^8 cm and a period of 11.74 sec, and the second, number 6, with a central temperature of 20.10^6 K has a radius of 9.72×10^8 cm and a period of 11.81 sec. When the temperature increases by 3 orders of magnitude,

TABLE II

Radial pulsation periods for white dwarfs

$M/M_\odot$	P_{sec} [1]	P_{sec} [2]	P_{sec} [3]	This work No.	P_{sec}	$\dfrac{\xi_s}{\xi_c}$
0.4	23.4	20.6	18.5	8	18.453	1.695
0.6	14.4	12.8	11.85	7	11.740	1.315
				6	11.811	1.500
0.8	9	9	8.3	5	8.167	1.220
1.0	6.2	6.4	5.9	4	5.740	1.060
				3	5.747	1.070
1.2	4	4.4	3.8	2	3.800	0.970
				1	3.797	0.970

Notes – Eigenfunctions of radial pulsations of white dwarfs. For the masses given in the first column, the table gives the periods from Harper and Rose (1969) (1) Ostriker and Tassoul (1969) (2) Cohen *et al.* (1969) (3) in comparison with our results. The last columns give the number of the models, their periods and the ratio of the amplitudes of the eigenfunction at the surface and at the center.

ite dwarfs

	10	11	12	13	14	15
cgs	ϱ_c gr	η_c	Non-isothermal shell (% of the radius)	partially degenerate shell %	H burning shell %	H content %
318×10^{25}	1.8719×10^8	6585	3.77	1.39	2	3.79×10^{-5}
915×10^{25}	1.8604×10^8	3398	4.08	3.1	1.77	2.0×10^{-5}
990×10^{24}	3.7042×10^7	1175	6.18	5.62	2.49	7.5×10^{-5}
389×10^{24}	3.7250×10^7	2185	5.15	2.67	2.95	9.38×10^{-5}
262×10^{23}	1.1318×10^7	501	9.35	8.88	3.06	1.46×10^{-4}
803×10^{23}	3.8218×10^6	228	15.2	14.22	3.87	3.45×10^{-4}
192×10^{23}	3.8759×10^6	705	8.8	4.99	4.66	4.85×10^{-4}
317×10^{22}	1.1957×10^6	266	15.7	12.32	6.8	1.2×10^{-3}

ar luminosity unit; 5: logarithm of the surface gravity; 6: logarithm of the effective temperature; 7: Bolometric
: value of the parametre of degeneracy at the center; 12: fraction of the radius occupied by the non-isothermal
the Hydrogen burning shell; 15: content in Hydrogen in fraction of total mass.

the radius increases by 8.5% and the period by 0.6%. This confirms the result of Hubbard and Wagner (1970) who find an increase of the period from 17.39 sec to 18.04 sec for a 0.43 $M_\odot$ model with a central temperature increasing from 0 K to 1.29×10^7 K (10^{-2} $L_\odot$). The more the mass of the models decreases, the more this effect of the temperature on the period is sensible. The increase of the period is 5.2×10^{-9} sec K^{-1} for our 0.6 $M_\odot$ models, and 5×10^{-8} sec K^{-1} for the Hubbard and Wagner's 0.43 $M_\odot$ models.

3. Vibrational Stability

The vibrational stability is studied by evaluating the overall dissipation according to the method described by Ledoux and Walraven (1958).

Three mechanisms can amplify the oscillations: (a) the classical x-mechanism in the external He II ionization zone; (b) the driving due to an external source of energy; here, the hydrogen burning shell; and (c) the increase of the perturbation in luminosity outward in the central part of the isothermal core due to the relative variation of the opacity and adiabatic gradient in this very degenerate region.

The relative influence of the different shells on the overall instability is shown by the behaviour of the energy integral throughout the star:

$$E = \int_0^M \left(\frac{\delta T}{T}\right)_a \delta \left(\varepsilon - \frac{1}{\varrho} \operatorname{div} F\right)_a dm$$

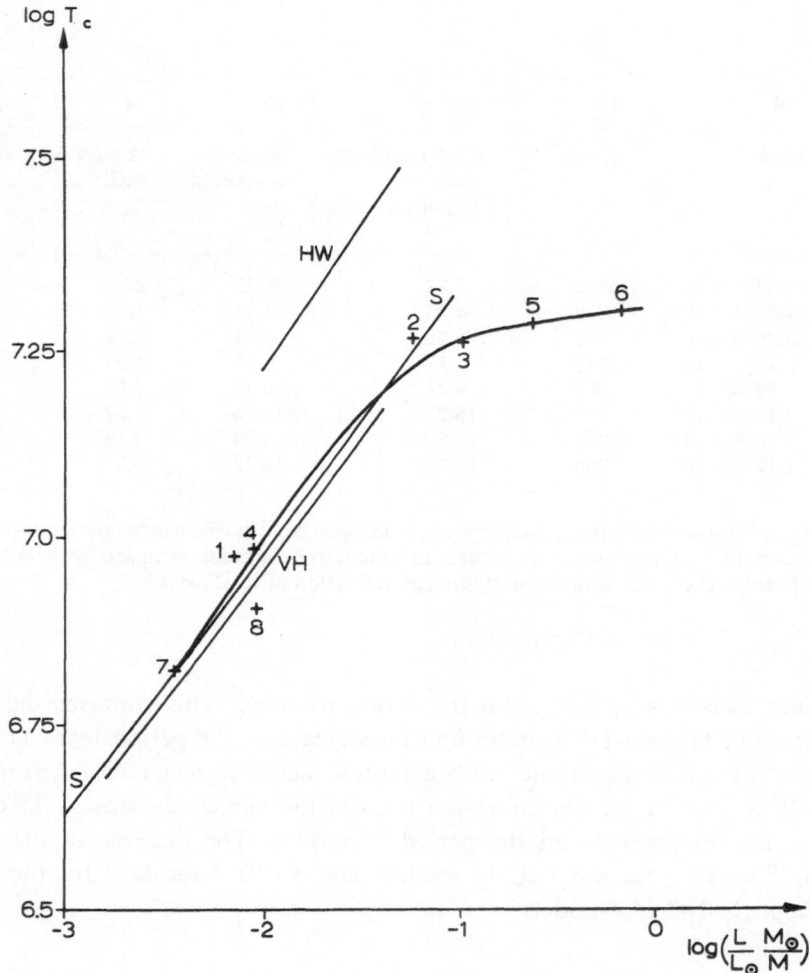

Fig. 1. Log T_c vs $\log(L/L_\odot \cdot M_\odot/M)$ for the eight models compared with Schatzman's (S), van Horn's (VH), and Hubbard-Wagner's (HW) relationships.

Figure 3 shows in model 1 the influence on the driving of the oscillations, due to the H and HeI ionization zone, and to the HeII one. The HeII ionization zone plays the most important role. At the edge of the isothermal core, the hydrogen burning shell provides a driving of the same order of magnitude as the HeII ionization one. The isothermal core itself begins to damp the pulsations but it amplifies them in its innermost central part. The x-mechanism alone is not able to trigger the instability. The behaviour of the energy integral in model 4 is analogous.

In model 8 (Figure 4) (and also model 7) the instability is mainly due to mechanism a. The excitation due to the H burning shell is indeed three times smaller than the excitation due to the ionization zone, and without this H burning shell the models would be unstable by x-mechanism alone.

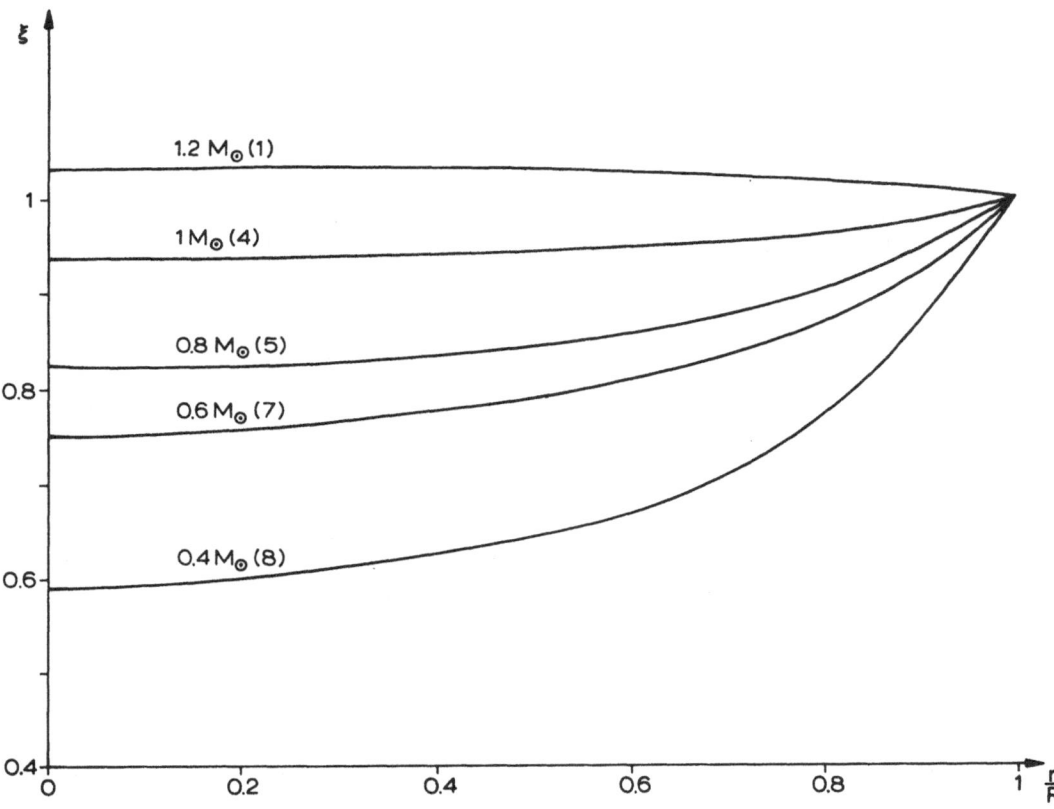

Fig. 2.   Eigenfunction in different mass models vs the relative radius.

TABLE III

No.	T (yr)
1	6.35 (10)
2	2.36 (8)
3	1.46 (8)
4	1.835 (10)
5	3.18 (7)
6	5.875 (6)
7	9.68 (9)
8	1.42 (9)

Note – In the first column: the number of the model; in the second one the amplification rate, the number in parenthesis is the power of ten.

On the contrary in model 5 (Figure 5) (and also models 2, 3 and 6) the amplification is caused predominantly by the mechanism b. The part of the isothermal core becomes negligible.

4. Remark

In this set of models of white dwarfs, the energy source considered is the hydrogen burning shell located at the edge of the isothermal core. In an improved treatment we should include the cooling of the degenerate core. But when this cooling is taken

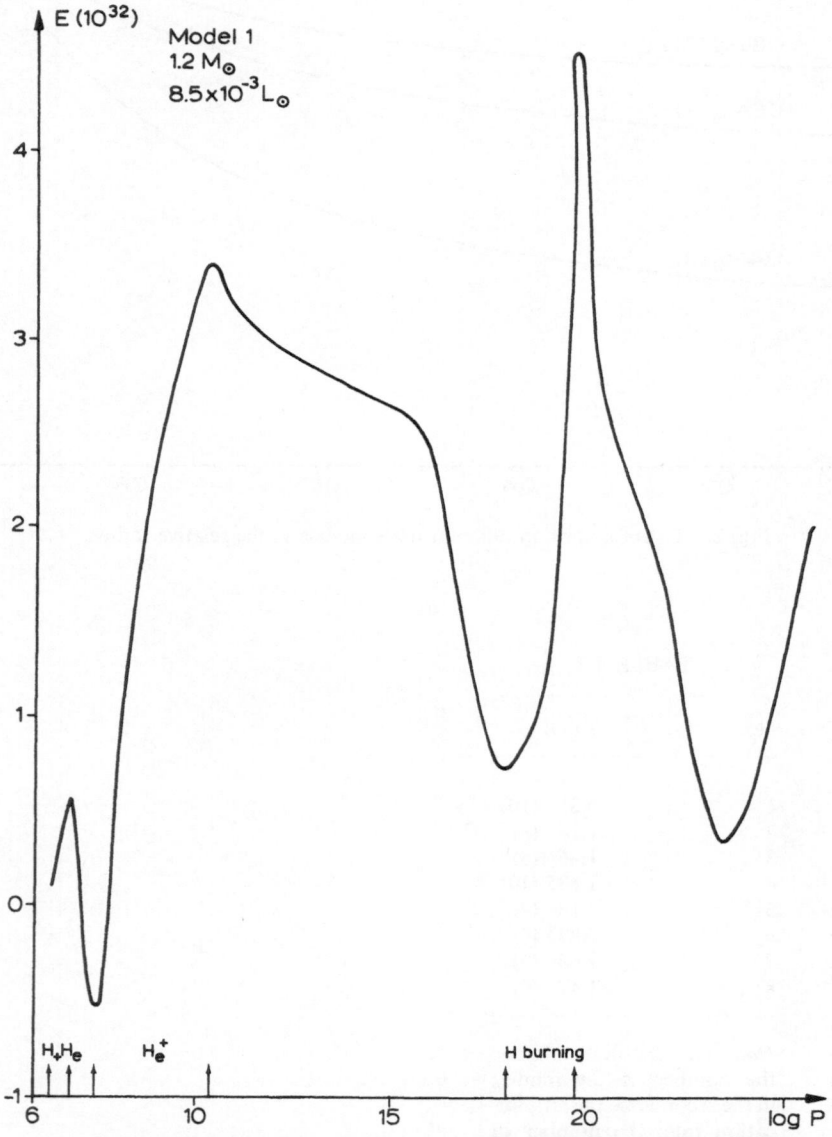

Fig. 3.   The energy integral vs the logarithm of the pressure in the model 1.

into account, the structure of the core is not modified because the electronic con-
ductibility in a degenerate gas keeps the temperature gradient very small such as we
may consider the core as isothermal.

For a model with a given luminosity, the contribution of the hydrogen burning
shell to the total luminosity would be smaller than for a model with only an H burning
shell. So, this shell would be thinner and removed outward.

But the fraction of the total mass occupied by the H burning shell is only in the

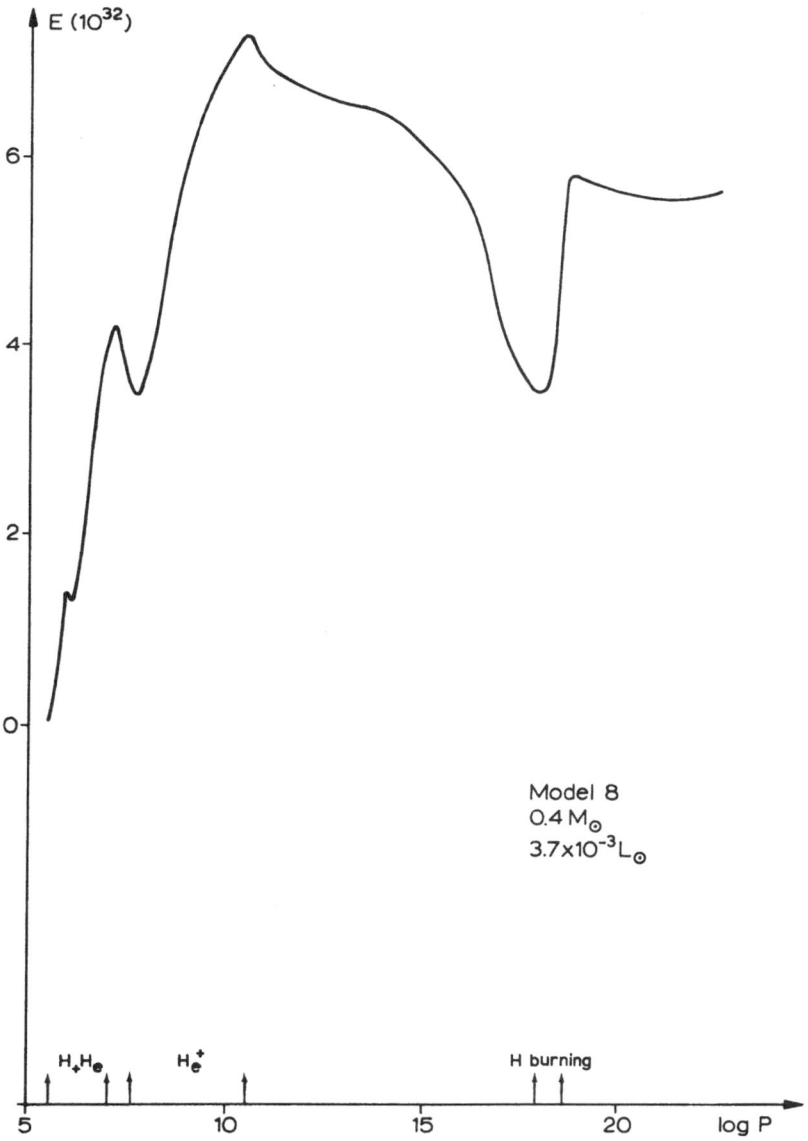

Fig. 4.   The energy integral vs the pressure in the model 8.

range of 10^{-4}–10^{-5}. The modification of both the locus and the width of this shell would not change the structure of the envelope which surrounds it. So the overall structure of the models would not be modified.

5. Conclusion

It could be expected that models of white dwarfs with an energy source located near the surface would be vibrationally unstable (Ledoux *et al.*, 1950; Baglin, 1967).

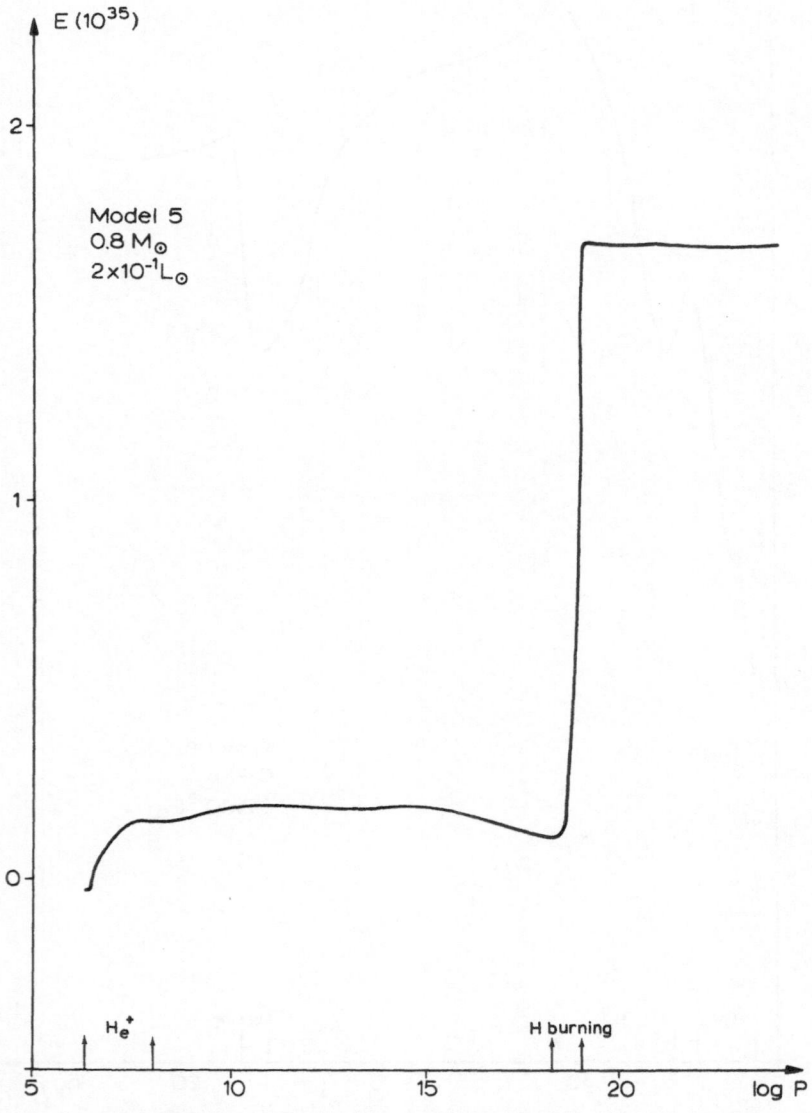

Fig. 5. The energy integral vs the logarithm of the pressure in the model 5.

Our preliminary results* show that, in fact, we find a region in the H − R diagram where the x-mechanism is mainly responsible for the instability.

It is interesting to notice that the two models which present this type of instability are located in the prolongation of the instability strip of the Cepheids (Christy, 1970) (Figure 6).

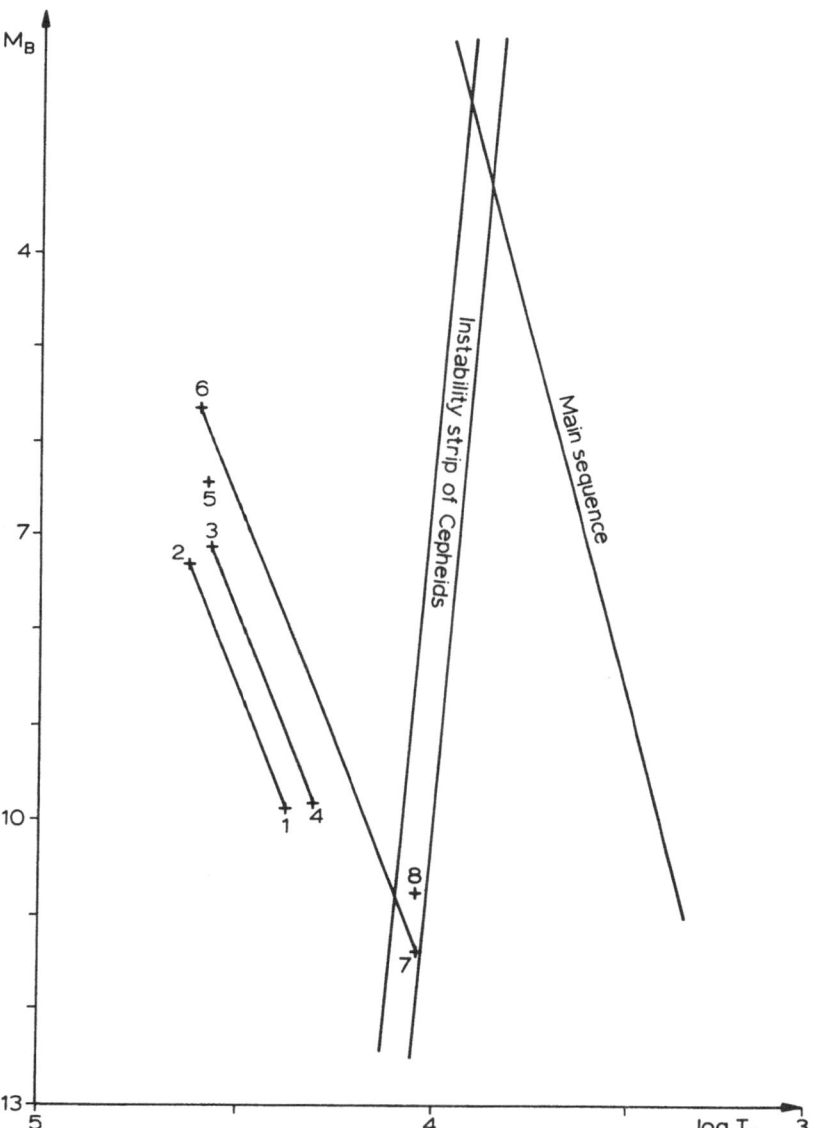

Fig. 6.   H − R diagram. Our models are located by a cross and their number as in Table I. The main sequence is given by Morton and Adams (1968) and the instability strip of the Cepheids by Christy (1970). Straight lines are drawn between models of same mass.

* More detailed results will be published later.

Variable white dwarfs can exist but they must be very difficult to observe. The amplitudes of the pulsations are very small; for a typical Cepheid, the amplitude of the luminosity variation is of the order of 1 mag; it decreases to $\frac{1}{100}$ of magnitude for δ Scuti variables; for white dwarf variables it must be less than $\frac{1}{100}$ of magnitude. On the other hand, the amplification rates found for our models are very long (Table III).

References

Baglin, A.: 1967, *Ann. Astrophys.* **30**, 617.
Chandrasekhar, S.: 1939, *An Introduction to Stellar Structure*, Ed. Dover.
Christy, R. F.:1970, *R.A.S.C.J.* **64**, 8.
Cohen, J. M., Lapidus, A. H., and Cameron, A. G. W.: 1969, *Astrophys. Space Sci.* **5**, 113.
Harper, R. R. and Rose, W. K.: 1969, preprint.
Hubbard, W. B. and Wagner, R. L.: 1970, *Astrophys. J.* **159**, 93.
Lauterborn, D.: 1970, *Astron. Astrophys.* **7**, 150.
Ledoux, P. J. and Sauvenier-Goffin, E.: 1950, *Astrophys. J.* **111**, 611.
Ledoux, P. J. and Walraven, T.: 1958, *Handbuch der Physik* **LI**, Springer-Verlag, Berlin.
Morton, D. C. and Adams, T. F.: *Astrophys. J.* **151**, 614.
Ostriker, J. P. and Tassoul, J. L.: 1969, *Astrophys. J.* **155**, 987.
Salpeter, E. E.: 1961, *Astrophys. J.* **134**, 669.
Schatzman, E.: 1952, *Ann. Astrophys.* **15**, 361.
Van Horn, H. M.: 1970, *Astrophys. J.* **160**, L53.
Vila, S. C.: 1966, *Astrophys. J.* **146**, 437.
Vila, S. C.: 1967, *Astrophys. J.* **149**, 613.

22. PROJECTIONS INTO THE FUTURE

F. ZWICKY

The Hale Observatories and California Institute of Technology
Pasadena, Calif., U.S.A.

1. The Specialized and the Morphological Approach

Science owes much of its success to the *specialized approach*, which endeavours to isolate objects and phenomena and to study them under elimination of the disturbing influences of the surroundings and internal contaminations. Thus, for instance, physicists grow pure single crystals of various chemical elements and compounds and study their mechanical, thermal, electrical, magnetic and optical properties under strictly controlled conditions of pressure, temperature and other parameters.

The specialized approach, however, involves several dangers. In the first place, through the investigation of isolated objects and circumstances one cannot be quite sure to arrive at significant results of general validity. And secondly, the specialized approach may lead to scientifically worthless and even non-sensical results in cases where the environment and disturbing influences are in reality impossible to avoid. Under such circumstances the use of the morphological approach * is indispensable. Finally, scientists practicing exclusively the specialized approach may become narrow minded specialists who fail to maintain a balanced outlook on all aspects of life, a trend which has resulted in the ever widening and humanly disastrous gap between science and the general public.

To mention one specific example to demonstrate the necessity and the usefulness of the general morphological approach as a tool in the special field of astrophysics which we are discussing in the present symposium, let us consider the problem of the evolution of a single star. Theoreticians, working on the internal constitution of stars and their development in time, generally have considered them as isolated from all external influences for hundreds and even thousands of millions of years, although neither they, nor any observer could possibly point to specific stars for which this is really true. If any cosmic object had actually passed unscathed through the billions of years we would not have any criteria for detecting or discovering it.

On the other hand, there cannot be any doubt that most of the stars as we see them today have been subject to more or less severe external disturbing effects, such as collisions with interstellar gas and dust clouds, exposure to both electromagnetic and corpuscular radiation, close encounters and the physical interaction with other stars,

---

* This approach, which has been gradually developed, perfected and applied in many fields during the past four decades, essentially concerns itself with the visualization, the analysis and the constructive use of all of the structural interrelations among objects, phenomena and concepts which might enter a given problem or set of circumstances. For those who might wish to try their hand at this approach a number of references are listed in the appended bibliography. [1]

Luyten (ed.), White Dwarfs, 155–164. All Rights Reserved.
Copyright © 1971 by the IAU.

as components of binary and multiple systems, etc. As a result of such interactions stars may accrete matter and increase in mass, or they may be whittled down.

For instance, if one component of a binary star becomes a supernova and blows off most of its mass, its companion, if not completely blown out of existence, that is disrupted and evaporated, will be more or less severely whittled down and become a 'run-away star' [2]. Depending on how much it has been whittled down it will fly away as a more or less normal star or, in the extreme as a very much smaller and hotter object, which I have proposed to call a ,pygmy star' of a density comparable to that of normal stars. About typical pygmy stars of very high density we shall speak later on. Whittled down pygmy stars are no doubt being produced in great numbers in many of the compact galaxies as well as in the compact cores and nuclei of certain spiral and elliptical galaxies, in which normal stars are being whittled down over more or less long periods of time in the intense interstellar radiation fields due to the emission of the other stars of the system.

As to systematic projections into the future, the morphological approach has developed certain means of scientific and technological forecasting, the absolute validity of which cannot really be strictly proved, but which, in the recent past, that is during the past four decades have been remarkably successful. Among the predictions made in the 1930's, and which have been confirmed since are, the existence of neutron stars and their relation to supernova outbursts and the generation of cosmic rays, the occurrence of faint blue stars of many types in open clusters and especially in high galactic latitudes (Humason-Zwicky stars), of gnome, pygmy and dwarf galaxies and, most recently, of the huge family of compact galaxies. The conclusion reached more than thirty years ago that the luminosity function of normal galaxies must be of the type proportional to $10^{0.2(M-M_0)}$, rather than to $10^{-a(M-M)^2}$ currently accepted at that time has now found its remarkably accurate confirmation.

The long standing predictions of compact galaxies acting as gravitational lenses and of the occurrence of nuclear goblins still need verification through more decisive observations than are available to us today. Likewise, the anticipated existence of various types of pygmy stars needs to be confirmed more convincingly.

Most of the above mentioned predictions were made by what I have proposed to call the *morphological method of directed intuition* supplemented by the idea that cosmic objects and phenomena will be found to be subject to the following general aspects.

(1) Objects will be found to exist in large and interrelated families.

(2) Any cosmic object can be formed slowly or fast, as measured in terms of speed of the motions and of the velocities of sound present in the original systems. For instance, in many heads the myth has established itself that neutron stars can only be formed fast, a myth which has prevented many theoreticians to see and evaluate cosmic evolutionary events in their true light.

(3) If time is long enough, that is much longer than the transit times between the particles or light quanta in a given system, then the principles of the Boltzmann-Gibbs statistical mechanics regulate the formation of compact and dispersed cosmic matter [3].

(4) And finally, it may reasonably be conjectured that to all fundamental lengths which can be formed from the basic physical constants, there exists always some type of characteristic matter or body. Concentrating in this short study mainly on this conjecture we may, among others, expect to find the following types of matter and of bodies.

It must of course be emphasized that these bodies will seldom be found in nature in pure form, but will always be 'contaminated' to a greater or lesser degree by types of matter belonging to other categories.

2. On the Hierarchies of Distinct Bodies

As mentioned above, it has in the recent past proved fruitful to relate the occurrence of different states of matter and of distinct bodies among the elementary particles, the atoms and molecules, microscopic and macroscopic solid and liquid bodies, as well as of a series of cosmic objects to the various types of basic lengths which can be expressed in terms of the known physical constants. We shall consider for this purpose the masses and charges m_p, m_e and $\pm e$ of the proton and the electron, the velocity of light c, the Planck constant h and the universal gravitational constant G. The masses of other elementary particles such as the mesons and hyperons will be mentioned in passing, while we can give only a furtive glance to the as yet not definitely known masses of neutrinos and gravitons.

It should also be noted that the basic lengths used below may in some cases be expanded to other significant lengths when multiplied with certain powers of two important *dimensionless numbers* which can be formed from the physical constants in the following way, namely

$$\alpha = 2\pi e^2/hc = \tfrac{1}{137} \qquad \text{(Fine Structure Constant)}$$

and

$$R = e^2/Gm_e^2 = 4.2 \times 10^{+42} \qquad \text{(Cosmic Number)}$$

The selection of basic lengths which we propose to choose is as follows.

$$d_B = h^2/4\pi^2 m_e\, e^2 = \text{BOHR'S LENGTH} = 5.3 \times 10^{-9}\,\text{cm}$$

As is well known, d_B is the characteristic length which essentially determines the sizes of free atoms and molecules. It is also important to remember that condensed matter in its usual macroscopic state, with a density of the order of unity (in terms of water at SPT) is generally electrically neutral in regions whose dimensions are of the order of d_B (in crystals the typical smallest electrically neutral components are the so-called elementary cells).

$$d_S = d_B/\alpha = 137\, d_B = 7.3 \times 10^{-7}\,\text{cm}$$

d_S might be called the atomic phase communication length. Since it limits the distance to which motions in crystals for instance can be strictly kept in phase, due to the finiteness of the speed of light, it also is related to the dimensions of a secondary structure even in the most perfect real crystals, most of which show dislocations and mosaic structure when the specimens have sizes of the order of one micron [1]. In this connection the interesting question arises how long the so-called whiskers of certain single crystals may actually become and if they might be formed in interstellar space in lengths impossible or very difficult to achieve on the earth, and what role they might play in causing obscuration and polarisation of light from more distant objects.

3. Interference of the Exclusion Principle

Before proceeding to the discussion of lengths smaller than d_B we must mention the relation of the Fermi statistics to states of matter different from those known to us in our immediate surroundings. Such states were first suggested by R. H. Fowler in order to explain the existence of the white dwarf stars with average densities of the order of 10^6 g/cm^3 and greater. Idealizing the model of an electronically degenerate star, all of the electrons, each one with its own specific set of quantum numbers must be thought of as occupying all of the lowest possible energy states within a matrix of protons or other nucleons as compressed by their mutual gravitational forces. This model, which would have the appearance of one giant molecule at the absolute zero of temperature must of course be modified by endowing it with some heat content and with a non-degenerate atmosphere subject to an equation of state of a more or less conventional gas.

With a view to some of the states of stellar matter to be discussed further on, whose interatomic distances are governed by the Compton wavelength λ_{Ce} of electrons, we may discover that this state closely overlaps the electronically degenerate state of white dwarfs discussed above, which alone seems to have been considered by most astrophysicists during the past four decades. Not having followed developments in this field very closely I am not aware of whether or not the relation of the states of matter in degenerate stars governed respectively by the exclusion principle and by the Compton wavelength λ_{Ce} for electrons has been discussed by the theoreticians and, in which way it is relevant for our understanding of the constitution of under-luminous dwarf stars, white and red, as well as possibly of dense pygmy stars.

$\lambda_{Ce} = h/cm_e = 2.43 \times 10^{-10}$ cm = COMPTON WAVELENGTH FOR ELECTRONS

This wavelength corresponds to a frequency v for which $hv = m_e c^2$, that is equal to the energy of complete annihilation of the electron. Therefore we may picture a state of matter associated with this Compton wavelength, which in the first place consists of a network of protons or of nucleons locked together at interparticle distances of the order of λ_{Ce}, and with virtual or real pairs of positive and negative electrons emerging between them. The minimum density of a star of this type and consisting entirely of hydrogen would be of the order of 10^5 g/cm^3. It should be added that the possibility of pairs of positive and negative electrons appearing and disappearing will add a new aspect to the problem of the stability or instability of stars of this type and account for some of the oscillatory features.

$\lambda_{Cp} = h/cm_p = 1.32 \times 10^{-13}$ cm = COMPTON WAVELENGTH FOR PROTONS

This wavelength corresponds to a frequency v, for which $hv = m_p c^2$, that is equal to the energy of the complete annihilation or irradiation of the proton. Therefore we may imagine a state of matter as being associated with this wavelength which, in the first place, consists of a network of protons, or of nucleons, locked together at interparticle distances λ_{Cp} by means of standing electromagnetic waves of this wavelength and, with virtual or real pairs of positive and negative protons, as well as with electrically neutral combinations of the lighter mesons and the positive and negative electrons emerging occasionally between the nucleons. The minimum density of a star of this type, and consisting entirely of hydrogen, is of the order of 8×10^{14} g/cm^3. This type of star, in addition to regular neutron stars, must therefore also be considered as a possible candidate for the compact remnants of certain supernovae and for pulsars.

Furthermore, since there are Compton wavelengths for all elementary particles, that is the heavier nucleons, as well as for the mesons and hyperons, both stable and unstable, we are actually confronted with the choice of a very great variety of possible degenerate stars, among which, as far as my own knowledge goes, no preference can immediately be given to one or the other type. These ideas, if confirmed, open up a distressingly large field for theoreticians and observers alike.

$d_{N1} = e^2/m_e c^2 = 2.8 \times 10^{-13}$ cm = FIRST NUCLEAR LENGTH

This length corresponds to the distance at which the positive electrical potential energy between two electrons is equal to the energy of annihilation of an electron, and also approximately equal to the energy of disintegration of a neutron into a proton and an electron. On the basis of arguments not to be repeated here, Baade and I in 1934 suggested the existence of neutron stars [4] and their probable relation to supernovae and cosmic rays. These hypotheses, which for more than thirty years found little or no credence among astronomers, seem now all of a sudden quite generally accepted, mainly as a consequence of the discovery of a pulsar as the stellar remnant of the supernova of 1054 AD which gave rise to the Crab nebula.

Here another thought must be inserted. It concerns the conjectured occurrence in the universe of degenerate dense bodies which are not of the thermodynamically stable (or strictly speaking pseudostable) type which we have discussed so far, but, which exist only temporarily because of local conditions of pressure, temperature, electric fields and so on and which are more or less highly explosive when released from these conditions. I wish to emphasize strongly that this idea promises to open up an enormous field of possibilities that is as yet unexplored but which will merit our serious attention if we are to understand the manifold as yet unexplained phenomena and events in the universe.

Here only one among these unstable compact bodies may be mentioned, namely the so-called *nuclear goblins**, some of whose characteristics I have discussed else-

---

* As designations of nuclear goblins in other languages I have suggested lutins nucléaires (Fr), Kernkobolde (G), folletti nucleari (I), duendes nucleares (Sp) and iadernie tshortiki (R).

where [5]. Nuclear goblins are supposed to be made up of the same type of nuclear matter as neutron stars for instance. With typical sizes of the order of one meter diameter and masses of the order of 10^{21} g they would release, when exploding in regions of low pressure or in free space, each, an energy of the order of 10^{39} erg. Goblins might exist in normal stars, or especially in white dwarfs when surrounded by pressures of the order of 10^{19} atm. When travelling accidentally towards the surface of a star or, when being expelled into interstellar space, they would explode and either cause flares in their hosts or appear as flashes in interstellar space. Scanning through tens of thousands of films and plates during my surveys for supernovae, clusters of galaxies and compact galaxies I have off and on encountered stellar images corresponding to apparent photographic magnitudes of the order of $m_p = 12$ or fainter and which, on plates taken 15 min later had disappeared, being then fainter than $m_p = 21$. I suggest that all observers who are engaged in similar survey work watch for such short duration flashes and that theoreticians consider the possibility that nuclear goblins escaping from the interior of certain stars causes them to flare, since exploding goblins can satisfactorily account for the amounts of energy released in such flares.

$d_{N2} = e^2/m_p c^2 = 1.6 \times 10^{-16}$ cm = SECOND NUCLEAR LENGTH

In contradistinction to d_{N1} we do not yet know what role this second nuclear length plays in the theory of the elementary particles of matter and the various nuclei of the atoms in the periodic system. If any type of condensed matter, in elementary particles or in condensed cosmic bodies were associated with it, the expected characteristic density of these objects would be of the order of 4×10^{23} g/cm^3. Such densities would assure any corpuscle of tremendous 'surface loading' and of a degree of penetrating power such as it is exhibited by the neutrinos. Whether or not there is any connection between d_{N2} and the values of the rest masses and densities of neutrinos probably no one can say at the present time. Equally obscure, of course remains the possibility that the ultimately collapsed objects in the universe might consist essentially of neutrinos, that is bodies which might have to be identified with what some physicists and astronomers call *black holes*, but which I personally prefer to designate as OBJECTS HADES, since they clearly cannot be holes but must be objects.

$d_0 = (Gh/c^3)^{1/2} = 4.05 \times 10^{-33}$ cm = COSMIC MINILENGTH

All that can be said at the present time about the significance of d_0 is speculative in the extreme, since we need more basic observational data, about neutrinos and gravitons for instance, as well as more profound knowledge about the nature of gravitation and of its interrelation with the electromagnetic field (unified field theory). Some suggestions may nevertheless be made which might inspire others to produce the missing links. These suggestions are firstly concerned with the following entirely nebulous conjectures 1 and 2, as well as with the somewhat more concrete ideas 3, 4 and 5.

1. *Discretization of Space*. Since the curvature of space and gravitation are related according to the general theory of relativity, a quantisation of the gravitational field

might lead to the discretization of space, subdividing it in some way into cells with dimensions of the order of d_0. It is not known to me whether or not this conjecture has as yet been formulated in any more quantitative way so as to allow predictions which might be checked observationally.

2. *Zero-Point Energy of Space.* Closely associated with the above suggestion is the conjecture that, to the gravitational radiation travelling through cosmic space there might correspond a permanent set of essentially standing waves of all wavelengths down to a value of the order of d_0, all of them representing the zero point energy content of space, analogous to the zero point oscillations in crystals or in a closed radiation space. Once we know the energy associated with any specific type of gravitational radiation the average mass density of the mentioned zero point energy could be calculated. It would of course be highly significant if it turned out to be greater than can be tolerated in any of the currently acceptable models of the universe and that therefore other causes than an actual expansion would have to be found for the universal redshifts in the spectra of distant galaxies.

3. *Sizes of the Neutrinos.* If d_0 is assumed to be characteristic for the size of the neutrino, this conjecture could be checked against our approximate knowledge of the penetrating power of neutrinos, provided that we succeed experimentally in determining their rest mass and provided that their velocity, or velocities, can be measured.

4. *Characteristics of the Gravitons.* If, in analogy to the quanta of light, the gravitons have a rest mass equal to zero, then both, the coulomb forces between two electric point charges and the gravitational forces between two masses, respectively, decrease indefinitely with the distance r between them like $1/r^2$.

Since, however, among ten thousand clusters of galaxies surveyed, there exist no clusters of clusters of galaxies and, in addition, the velocity dispersion among neighboring clusters is only of the order of a few thousand kilometers per second instead of the expected ten thousands of kilometers per second, the simplest explanation of these facts is that at indicative distances of about $\Lambda = 20 \times 10^6$ parsecs from any given mass its gravitational field declines more rapidly than $1/r^2$. Taking Λ as the Yukawa length characteristic for the gravitational forces, it follows that the gravitons, as the exchange particles corresponding to these forces, have masses of the order [6]

$$m_G = h/c\Lambda = 5.65 \times 10^{-64}\,\text{g}.$$

About the same values for Λ and m_G can be derived independently from the hypothesis that the redshift in the spectra of distant galaxies is due to the gravitational drag of light [7], a coincidence which may be viewed as numerological in nature, as long as no additional confirmations are found.

Conjecturing in addition that the diameter of the gravitons might be of the order of d_0, they then would have to be assigned a mass density of the order of 2×10^{34} g/cm^3, which would be ample to explain their immense penetrating power, but which still would leave open the possibility of shielding any gravitational field to some degree by interposing some very dense objects, such as nucleons, neutron and hyperon stars, as well as of course objects 'Hades'.

5. *Object HADES – the Ultimate Collapsed Configuration.* We assume that matter, via white dwarfs, neutron stars and other degenerate highly condensed configurations may collapse only to the limit at which the energy lost on the way, divided by c^2 becomes equal to the initial 'dispersed' rest mass m_0 that is involved. We then have for the limiting final mass, both inertial and gravitational,

$$m_f = m_0 - \text{energy lost}/c^2 = 0$$

I propose the designation OBJECT HADES for such an ultimately collapsed configuration, because nothing can get away from it. In reality, when surrounded by other matter and radiation, m_f can never quite become equal to zero, and objects HADES will have some rudimentary 'thermal life' left in them which might correspond to a state of perhaps a few degrees Kelvin above absolute zero.

As to the internal composition and constitution, it may be conjectured among other possibilities, that object HADES is a cosmic ball of 'standing electromagnetic waves' which hold each other in place by their quantized masses $hv/c^2 = h/c\lambda$, where $\lambda_{\min}$ might be possibly of the order of the cosmic minilength d_0, and thus the maximum quantum in the zero point radiation complex would be $hv_0 = 5 \times 10^{16}$ erg.

At this perfectly enormous maximum energy of the light quanta, these may of course be expected to violate the superposition principle and the resulting interactions between them will endow object HADES with an inhomogeneous mixture of a multitude of neutral and of charged particles, such as gravitons, leptons and baryons. Just what the composition of this mixture, as well as its energy spectrum and that of the zero point electromagnetic radiation might be must obviously be left to a future analysis, which can be attempted successfully only after we have gained much more knowledge on the nature of the gravitons, the neutrinos, mesons and all of the other elementary particles as well as the physical laws which govern their interactions.

As to a possible observational search for objects HADES I have long ago suggested [8] the use of full size objective gratings mounted on large Schmidt telescopes and also inserted in the parallel beams between the components of zero correctors of large reflectors. What one may expect to see in the case of dead neutron stars and objects HADES is of course not their own light but light from surrounding sources bent around them towards the observer (see gravitational lens effects [9]).

4. Miscellaneous suggestions Concerning Extended
Electric Charges and Fields

Magnetic fields in stars and galaxies have been considered ad infinitum in astrophysics, although little or nothing fundamental has been written about their origin. I hope to discuss some relevant ideas about this subject in another place.

The possible occurrence and the analysis of extended distributions of electric charges and fields, on the other hand seems to have been largely neglected, although they no doubt play an important role in some cases. For the reasons indicated in the following, more or less extended regions in both cold and very hot bodies may carry net charges which may materially influence the internal characteristics of these bodies and their interactions with other cosmic bodies.

In hot stars, for instance, large numbers of positive and negative electric charges may get separated as a consequence of the interplay of fluctuations in the heat content and the electrical potential energy. I think it can be shown that this separation of electrical charges, which occurs in all dimensions, starting from interatomic or interionic distances to the existence of net charges on almost any two hemispheres of a star may in some cases require significant alterations of the equations of state which so far have been overlooked or disregarded.

Net total electric charges can clearly be carried by solids, starting from interstellar dust particles up to meteors, asteroids, small planets and bodies of the size of planets and possibly greater.

The fundamental question immediately arises as to the sizes up to which conducting, respectively no-conducting insulating spheres can carry charges, such that the coulomb forces between them are greater or equal to their mutual gravitational attraction. Disregarding here insulating bodies, that is the possibility of having a net charge frozen in their interior, we consider only solid spheres carrying a uniformly distributed electric charge on their surface. Assuming that cold electron emission would set in if the sphere were charged up to 10^9 V, the corresponding critical size of the spheres, at an average density of 1 g/cm^3 would be about 1 km.

If therefore no interstellar plasma, compensating the net charge, is carried along by our 'planetoids' of 1 km diameter their electrical and gravitational interactions would be of the same order of magnitude.

Any tests of the general theory of relativity, using observations of the orbits of freely flying projectiles launched from the earth, or of planetoids like Hermes, Icarus, Halleria, Berna and Glarona, whose diameters range from 100 m to about 20 km, are therefore subject to severe objections. Even the interpretation of the perihelion motion of Mercury would seem to be in doubt if variable net charges are considered as well as line-up of electric forces differing in time and space from those due to gravitation.

As to continuing the search for as yet unknown types of cosmic objects, a discussion comes to my mind which, about thirty years ago I had with Professor C. G. Darwin. Driving him from Pasadena to the Palomar Observatory, just after having discovered a number of supernovae and the Humason-Zwicky stars, I suggested in my enthusiasm that probably one new cosmic object could be discovered every day for many years to come. Darwin responded in his typically British manner "Then, why don't you?".

Considering the number of supernovae, peculiar faint blue stars and possible pygmy stars, the pulsars and neutron stars, the gnome, pygmy and dwarf galaxies, the compact galaxies, the radio sources and quasars and the intergalactic matter discovered since, my estimate in 1939 was not too far off the mark.

If the generation following us is still of a mind of trying to discover one new cosmic object every day, Darwin's remark may accompany them on their way "Then, why don't you?".

References

[1] Zwicky, F.: 1957, *Morphological Astronomy*, Springer-Verlag, New York.
   Zwicky, F.: 1961, *Morphology of Propulsive Power*, Monograph No. 1, of the Society for Morphological Research, Pasadena.
   Zwicky, F.: 1969, *Discovery, Invention, Research*, Macmillan.
   Zwicky, F. and Wilson, A. G.: 1968, *New Methods of Thought and Procedure*, Springer-Verlag, New York.
[2] Zwicky, F.: 1953, *Publ. Astron. Soc. Pacific* **65**, 205.
[3] Zwicky, F.: *Astron. J.* **69**, 565 (1964); *Astrophys. J.* **140**, 1467 (1964); *Acta Astron. Warsaw* **14**, 151 (1964).
[4] Baade, W. and Zwicky, F.: *Proc. Nat. Acad. Sci.* **20**, 254 (1934); and **20**, 259 (1934).
[5] Zwicky, F.: 1958, *Max Planck Festschrift*, Berlin, p. 243.
[6] Zwicky, F.: 1961, *Publ. Astron. Soc. Pacific* **73**, 314.
[7] Zwicky, F.: 1929, *Proc. Nat. Acad. Sci.* **15**, 773.
[8] Zwicky, F.: 1941, *Phys. Rev.* **59**, 221.
[9] Zwicky, F.: 1937, *Phys. Rev.* **51**, 290; and **51**, 679.